THE HISTORY OF CHEMICAL TECHNOLOGY

BIBLIOGRAPHIES OF THE HISTORY
OF SCIENCE AND TECHNOLOGY
(Vol. 5)

GARLAND REFERENCE LIBRARY
OF THE HUMANITIES
(Vol. 348)

Volume 5

Bibliographies of the History of Science and Technology

Editors

Robert Multhauf
Smithsonian Institution, Washington, D.C.

Ellen Wells
Smithsonian Institution, Washington, D.C.

THE HISTORY OF CHEMICAL TECHNOLOGY
An Annotated Bibliography

Robert P. Multhauf

GARLAND PUBLISHING, INC. • NEW YORK & LONDON
1984

© 1984 Robert P. Multhauf
All rights reserved

Library of Congress Cataloging in Publication Data
Multhauf, Robert P.
  The history of chemical technology.

  (Bibliographies of the history of science and technology : v. 5) (Garland reference library of the humanities : v. 348)
  Includes index.
  1. Chemistry, Technical—History—Bibliography.
I. Title. II. Series. III. Series: Garland reference library of the humanities : v. 348.
Z7914.C4M84 1984 [TP15]   016.66′09   82-48272
ISBN 0-8240-9255-4 (alk. paper)

Cover design by Laurence Walczak

Printed on acid-free, 250-year-life paper
Manufactured in the United States of America

GENERAL INTRODUCTION

This bibliography is one of a series designed to guide the reader into the history of science and technology. Anyone interested in any of the components of this vast subject area is part of our intended audience, not only the student, but also the scientist interested in the history of his own field (or faced with the necessity of writing an "historical introduction") and the historian, amateur or professional. The latter will not find the bibliographies "exhaustive," although in some fields he may find them the only existing bibliographies. He will in any case not find one of those endless lists in which the important is lumped with the trivial, but rather a "critical" bibliography, largely annotated, and indexed to lead the reader quickly to the most important (or only existing) literature.

Inasmuch as everyone treasures bibliographies, it is surprising how few there are in this field. George Sarton's *Guide to the History of Science* (Waltham, Mass., 1952; 316 pp.), Eugene S. Ferguson's *Bibliography of the History of Technology* (Cambridge, Mass., 1968; 347 pp.), François Russo's *Histoire des Sciences et des Techniques. Bibliographie* (Paris, 2nd ed., 1969; 214 pp.) are justifiably treasured but they are, of necessity, limited in their coverage and need to be updated.

For various reasons, mostly bad, the average scholar prefers adding to the literature to sorting it out. The editors are indebted to the scholars represented in this series for their willingness to expend the time and effort required to pursue the latter objective. Our aim, and that of the publisher, has been to give the series enough uniformity to give some consistency to the series, but otherwise to leave the format and contents to the author/compiler. We have urged that introductions be used for essays on "the state of the field," and that selectivity be exercised to limit the length of each volume. Since the historical literature ranged from very large (e.g., medicine) to very small (e.g., chemical technology), some bibliographies will be limited to truly important writings while others will include modest "contributions" and even primary sources. The problem is intelligible guidance into a particular field—or subfield—and its solution is largely left to the author/compiler. In general, topical volumes (e.g., chemistry) will deal with the subject since about 1700, leaving earlier literature to area or chronological volumes (e.g., medieval science); but here, too, the volumes will vary according to the judgment of the author. The volumes are international (except for two, *Science and Technology in the United States* and *Science and Technology in Eastern Asia* but the literature covered depends, of course, on the linguistic equipment of the author and his access to "exotic" literatures.

Robert Multhauf
Ellen Wells

Smithsonian Institution
Washington, D.C.

CONTENTS

Introduction	xi
Abbreviations	xvii
List of Illustrations	xix

PART I

Text and Reference Books	3
Reports on International Exhibitions	22
General Historical Literature	28
National Histories	
Asia	39
Austria, Hungary and Czechoslovakia	45
France	48
Germany	51
Great Britain	58
Holland and Belgium	64
Italy	66
Russia and Poland	67
Spain, Portugal, and Latin American	69
Switzerland	70
United States and Canada	71
Biographies	79
Company Histories	86
Chemical Hazards	99

PART II

Acids and Alkalis	108
Alum and Vitriol	114
Borax	119

Ceramics 123
Colors and Dyes 127
Food and Drink 137
Glass 144
Leather 151
Lime, Gypsum and Cement 152
Organic Products 155
Paper 156
Pyrotechnics 159
Rubber and Natural Plastics 161
Sal Ammoniac 164
Salt 166
Saltpeter and Chili Saltpeter 179
Soap and Detergents 185
Sugar 186

PART III

Abrasives 189
Agricultural Chemistry 189
Apparatus 194
Artificial Fibres and Plastics 196
Bleaching 202
Coal 204
Dyes, Synthetic 205
Electrochemical Industries 211
Explosives 213
Fuels, Exotic 215
Gas, Heating and Illuminating 216
Gases, Industrial 219
Inorganic Synthesis 223
Nitrates 224
Organic Synthesis 226
Petroleum 227

Contents

Pharmaceuticals	229
Photography	236
Potash	241
Soda, Artificial	244
Sulphuric Acid	250
Textile Chemistry	252
Author Index	253
Title Index	284

INTRODUCTION

As a rule, the ancients discussed technology, when they mentioned it at all, in terms of the "mechanic arts," under which rubric they included the whole subject. A serious attempt to classify these arts began in medieval Europe and led, by the eighteenth century, to the habit of subdividing them into mechanical and chemical, giving the two more or less equal weight. By the later eighteenth century the chemical arts were often further subdivided into those whose processes were "wet" (involving solutions) and those which were "dry" (involving heating). This development is revealed in the descriptions of early text and reference books (Part I).

The relative parity between the mechanical and chemical arts in early literature is not reflected in the modern literature of the history of these two topics, for there is an enormous disproportion, both quantitatively and qualitatively, in favor of the former. This undoubtedly owes much to the construction of the idea of "the industrial revolution" around innovations in machines and sources of power, but it also owes something to the understandably greater appeal of steam engines, locomotives and other mechanical relics to anything that can be recovered from the chemical industries. The result, as embodied in modern histories of technology, gives an impression of technology which is unbalanced and is contradicted both by the early literature on that subject and by the technology which surrounds us today.

Old textbooks further show that "chemical technology" was until very recently a generic term for semi-independent "arts" such as glass-making, ceramics, pigments and dyes, fuels, rubber, etc., and included only a few topics with much direct relationship to the *science* of chemistry. The manufacture of sulphuric acid and of artificial soda (sodium carbonate) were the most conspicious of these, and even in these cases the subject was empirical to the degree that the involvement of science is questionable. The chemist was willing enough in the eighteenth century to apply his science, but he was not very successful. The proprietors of nineteenth century technology saw little reason to credit science for any part of the great engine they had created, while the scientist tended to feel himself well rid of the responsibility and cultivated "pure" science.

But in the last half of the nineteenth century, Germany, long the home of the largest number of chemists and of the weakest chemical industry among the industrialized nations, found use for the former and a remedy for the latter in two areas of the traditional chemical industries, potash and dyes. The discovery of subterranean beds of mixed minerals, including potassium salts (c. 1850), presented opportunities and problems beyond the ken of the traditional technologist. So did the discovery of William Perkin (1856) that dyes could be made by manipulating the complex ingredients of coal tar. With an abundance of chemists and an industry which had little to lose, Germany proceeded to construct a whole spectrum of new chemical technologies. By 1900 there were important chemical industries more closely related to science than to the chemical arts of tradition; and they were controlled by chemists.

The twentieth century has seen a general reconstruction of chemical industries after the German model. Innovation, no longer restricted to Germany, has created a science-oriented "chemical engineering" and an industry of enormous variety which is nevertheless unified around the concept of "unit operations," a scientific refinement of the old idea that processes could be wet or dry. Today it is the empiricist, not the scientist, who is the outsider, for the former is largely restricted to the surviving fragments of the traditional industries, glass-making, rubber, fuels and the like—industries which are themselves increasingly science oriented.

All this is very recent, and an historical bibliography of the new industrial chemistry would be a very short one. Not only is the time-span short but the proprietors of the science-based industry have shown themselves (with the exception of those who actually founded it, as will be seen below) to be as indifferent to their own history as were the empiricist proprietors of the nine-

Introduction   xiii

teenth century. The future is more exciting, and it is assumed that prosperity—generally called "progress"—will last forever—or at least until the present proprietors are out of the industry.

The historical bibliography of the traditional industries is larger, in part, because their often moribund state makes them convenient targets for the professional historian. The bulk of the historical literature on traditional technologies also stems from their importance in economic history and/or the history of art. It is considerably reduced when they are considered as chemical technologies. The principal value, for example, of including such a topic as glass in this bibliography is perhaps the focus on technology, which is usually difficult to extract from the numerous and voluminous bibliographies on the history of glass.

These circumstances have made appropriate a subdivision of this bibliography into "traditional" and "modern" chemical technologies, and such a subdivision has been made. Like all subdivisions, it is imperfect. "Traditional" does not here equal "backward," and it has seemed best to include all entries for some technologies (e.g., glass and ceramics) under traditional industries. I have also followed this practice in some cases where it is less justified (e.g., rubber) where it seems less inconvenient than any attempt to split the industry into traditional and modern components. Some technologies do seem to divide naturally into traditional and modern (e.g., dyes, acids, and alkalis) and they are so divided. For the rest, sugar is a traditional technology, coal is a modern technology—the classification in these cases being determined by the bulk of the relevant historical literature.

Perfect segregation of topics is indeed rendered impossible at the outset by the existence of a general historical literature which may include any or all of the special topics. This general literature, here included under Part I, is subdivided further through the inclusion of "national" histories, the histories of firms, and biographies—to which I have appended a section on "chemical hazards," which seems not inappropriate. All of these subdivisions are naturally intended to make the book easier to use, but the reader who seeks everything on a particular topic must have recourse to the index.

Both quantitatively and qualitatively, chemical technology is among the most neglected areas of the history of technology.

Relatively little has been written on it, and of that which has appeared a large part has a celebratory or commemorative character'which can only by an excess of courtesy be called history. Since this is a selective bibliography I have tried to eliminate publications of small importance, but, alas, they are often all that exists. The annotation should assist the reader in determining whether any particular publication is worth the trouble to find.

This weakness is particularly true of histories of the modern industry. But there are exceptions, notably the industries of artificial soda, synthetic dyes, potash and synthetic ammonia. The first of these has been picked up (with good reason) by historians of the industrial revolution (or revolutions) as the favored example of the involvement of chemistry; and the result has been a large and competent historical literature. The size and quality of the literature of the other three examples seem to stem from their obvious importance in the genesis of the modern industry, and perhaps even more from their origins in Germany, where serious history seems to have remained fashionable among the practitioners of science and technology longer than elsewhere.

Because of the weakness of the historical literature I have begun most sections with "early works," generally what might be called "primary sources"—listed chronologically—which at least give some idea of the point at which a history might begin. The most elaborate of these is the first, "text and reference works," for writings of this kind were long more numerous than special works, and often preceded them. These early publications, moreover, are not only by far the fullest sources on early stages of the art but are often replete with historical information.

This bibliography is intended to be international but is regrettably deficient in many respects. English, French and German literature predominate in the libraries I have used and I have been unable to use literature in the Slavic and Oriental languages. Another deficiency is the limited extent to which I have recovered historical materials published in the professional (not historical) periodical literature. My excuse is the difficulty of finding them. There is an imperfect correlation between volumes and years, there are "supplements," often unfindable, and there are changes in title and other peculiarities (the *Zeitschrift für angewandte Chemie* only began numbering its volumes in 1905,

Introduction

with vol. 18!). Professional journals distribute advertising pages ingeniously and fail almost universally to include the word "history" in indexes. The task of locating historical articles, in short, is analagous to the proverbial search for a needle in a haystack. Most of the historical articles from professional journals included here were discovered because some previous historian had referred to them.

A number of the items listed here are rare. This hardly seems sufficient reason not to mention them, and indeed modern interlibrary loan and photocopying go far to eliminate a researcher's excuse for failing to consult a work on the ground of unavailability. But is it worth the trouble to pursue a particular publication? The annotations are intended to assist in answering that not insignificant question. The reader will notice that I have not myself seen a considerable number, perhaps 10%, of the entries. Where I found no evidence, beyond my initial reference, that a book or article actually exists, I have eliminated it. Books included but "not seen" are in most cases simply not available in the libraries I have used, which were principally the Library of Congress, the Smithsonian Institution and (on short visits) The New York Public Library, the libraries of Harvard University, and several European libraries. The time and energy available to me have not permitted resort to interlibrary loan. Some items were not seen simply because they were missing from library shelves for extended periods.

Some estimate of the adequacy of coverage of specific topics may be gained from consideration of the section, "salt," since I have written a book on this topic and have thus presumably covered it better than others. And indeed, with 83 entries, "salt" has the largest number in the bibliography. On consideration, however, I doubt that the number is disproportionate. Salt is probably the oldest, most widespread and even today the most important "industrial chemical." I doubt that the space here given to salt indicates a relative neglect of other topics—and yet, for the history of another substance of great antiquity and industrial importance, sulphur, I have found scarcely anything!

Certain bibliographical conventions have been violated for the following reasons. I have only included the names of publishers for very recent books or where the publisher is unfamiliar. It is always possible to find a publisher's name where the author or title is known but hardly ever the reverse. On the other hand, I have included the number of pages in books (except in a few cases where unavailable), for this is surely an important consideration to the potential reader. As regards "citations," I have tried to use this term only where there are citations to genuine sources, not, as often, vague allusions to whole books or even secondary works. Useful references are listed as "references." But it is impossible to be very exact in this matter. Bibliographies are sometimes more useful than citations, and works without citations often have useful references buried in the text. Where I have made no mention of this aspect of a book or article, it generally signifies an inability to decide what to make of it.

ABBREVIATIONS

AdS	Academy of Sciences/Académie des Sciences/ Akademie der Wissenschaften. In the case of AdS Stockholm, my reference is to the German translation (Königliche Schwedischen Academie der Wissenschaften, *Abhandlung*, Hamburg, 1739–83.
AGNT	*Archiv für Geschichte der Naturwissenschaften und der Technik*
AIChE	American Institute of Chemical Engineers
APS	American Philosophical Society
AS	*Annals of Science*
BAAS	British Association for the Advancement of Science
BASF	Badische Anilin und Soda-Fabrik
BSE	*Bulletin de la Société d'Encouragement pour l'Industrie Nationale*
CIBA	Gesellschaft für chemische Industrie in Basel
CR	*Comptes Rendus*, Académie des Sciences, Paris
DHA	*Documenta aus Hochster Archiv*
DCG	Deutsche chemische Gesellschaft
GGBB	Gesellschaft für die Geschichte und Bibliographie des Brauwesens
ICI	Imperial Chemical Industries
I&EC	*Industrial and Engineering Chemistry*
JCE	*Journal of Chemical Education*
JFI	*Journal of the Franklin Institute*
NN	*Naturhistorischer und Chemisch-Technische Notizen*
NTM	*NTM: Zeitschrift für Geschichte der Naturwissenschaft, Technik, und Medizin*
PT	*Philosophical Transactions of the Royal Society of London*
SB	*Schriftenreihe der Unternehmensarchivs der BASF*

SIAR	Smithsonian Institution, *Annual Report*
TNS	*Transactions of the Newcomen Society*
T&C	*Technology and Culture*
ZAC	Zeitschrift für angewandte Chemie
ZBHS	Zeitschrift für Berg-, Hütten-, und Salinenwesen im preussischen Staate

LIST OF ILLUSTRATIONS

Figure 1. Saltpeter production. From M.B. Valentine *Museum museorum*, Frankfort/M, 1704-14.   2

Figure 2. Apparatus for distillation of iron vitriol to produce "Nordhausen" (fuming) sulphuric acid. From Figuier (item 112).   after page 108

Figure 3. Interior of the dye works of Les Gobelins, France. From Turgan (item 226).   after page 126

Figure 4. Beaters and washers at the paper factory of Essone, France. From Turgan (item 226).   after page 156

Figure 5. Sea salt production near Croissic, Brittany. From Figuier (item 112).   after page 166

Figure 6. Cane sugar factory, Cuba. From Figuier (item 112) after page 194

Figure 7. Eduard Adams apparatus for the distillation of alcohol, 1801. From *Dingler's Polytechnisches Journal*, Vol. 2, 1820.   after page 194

Figure 8. Electric battery, plating tank, and electrical connections of the galvanoplastic works at Auteuil, France. From Turgan (item 226).   after page 210

Figure 9. Battery of retorts used in the production of gas for heating and illumination. From Turgan (item 226).   after page 216

Figure 10. Sectional view of salt and potash deposits at Stassfurt. From Figuier (item 112).   after page 240

Figure 11. Solvay process for soda, as shown on the United States patent of 1873.   after page 240

Figure 12. Cross-section of chamber sulphuric acid works, with (left) a Gay-Lussac tower. From Figuier (item 112).   after page 250

The History of
Chemical Technology

Figure 1. Saltpeter production. From M.B. Valentine *Museum museorum*, Frankfort/M,

PART I

TEXT AND REFERENCE BOOKS

As noted in the introduction, these works are included because textbooks are often valuable as "primary" sources, and because--up to the later 19th century--they typically go into the history of the topics treated. These topics, moreover, define "chemical technology," which has not always been what the modern chemist thinks it is, a matter not irrelevant to the history of the subject.

1. BIRINGUCCIO, Vannuccio. Pirotechnia. Venice, 1540.
   (158 pp. English trans. by C.S. Smith & M.T. Gnudi, New York, 1943)
   A pioneering work by a Sienese engineer of many talents, primarily on metallurgy but including such "semi-minerals" as sulphur, vitriol, alum, salt, borax and glass, as well as "various works of fire," such as ceramics, bricks and lime, and "artificial combustible materials," that is, saltpeter, gunpowder and fireworks. The better known De re metallica of Georg Agricola (1556) covers nearly the same subjects. He took much from Biringuccio and improved on him, but not much. Biringuccio wrote in Italian, Agricola in Latin.

2. GLAUBER, J.R. Furni novi philosophici, sive descripto artis destillatoriae. Amsterdam, 1648. (5 parts.

English trans. by Christopher Packe, 1689)
Here the traditional medicinal recipe book became a significant guide to the chemicals of commerce. Chemicals which were not distillation products were treated in others of Glauber's numerous books, such as his Operis mineralis (1651, often bound with the "Furnaces"). Glauber was himself a chemical manufacturer in Amsterdam, the contemporary center of the industry, and he was not very secretive. Hence, this book is a mine of information; but the reader has to dig for it.

3. SHAW, Peter. Chemical lectures for the improvement of arts, trades and natural philosophy. London, n.d. (1732 and later editions). (about 470 pp.)
His "aim" is as indicated by the title, but he claims that "a more philosophical chemistry" will benefit philosophers as well as "men of business." Hence the work is somewhat attenuated as far as chemical technology is concerned. The industrial topics treated more than cursorily are, wines and spirits, colors (dyes), pharmaceuticals, pyrotechnics and metallurgy.

4. ZEDLER, J.H. Grosses vollstanändiges Universal-Lexikon. Halle, 1732-50. (64 vols)
This encyclopaedia (of which Zedler himself wrote only the first 19 volumes) is earlier and larger than the more famous Diderot (no. 8), but smaller than the Krünitz encyclopaedia (no. 16). All have comprehensive articles on most of the traditional products of chemical technology. Saltpeter (which I will frequently take as an example) covers 42 columns, more or less equivalent to that many pages. References are commonly included.

5. BOERHAAVE, Hermann. Elementa chemiae. Leyden, 1732.
There are several editions, indeed, several versions, including the English translation, New Method of Chemistry, London, 1741 (2 vols with notes by Peter Shaw).
This landmark among chemical textbooks is divided into three parts, the first on "the origin, progress, cultivation and fortune" of chemistry, the second on "the principles of chemistry," and the third, "practical part," on "the actual operations of chemistry." Part three occupies the second volume (410 pages), and

deals with 227 "processes," mostly for the preparation of medicines, which at that time meant virtually anything and hence included the common industrial chemicals. There is little indication that any chemical was produced in quantity. Shaw's profuse notes to the English edition deal more extensively with the production and use of common commercially important materials.

6. ZINCKE, G.W. Grundriss einer Einleitung zu den Kameralwissenschaft. Leipzig, 1742. (2 vols)
This contains the earliest classification of chemical industries known to me, under the subtitle Schmelz und Sied-Werks Geschäfte (The business of melting and boiling works). They are (1) smelting, as in metal production, (2) boiling, as of salt production from brine, (3) saltpeter boiling and working, and (4) other small burning, melting and boiling operations, such as dye works, lime burning, glass furnaces, mirror making, brick and pottery ovens and porcelain works. He gives some detail on the principal industries, especially salt and saltpeter, which were state monopolies. In the German states the cameralists (practitioners of Kameralwissenschaft) correspond roughly to the English Privy Councilors, but in these centralized states their range of activity was far wider, and included technology, in which they were often remarkably well-informed. Zincke is the first of the several who will be included here.

7. MACQUER, P.J. Eleméns de chymie practique. Paris, 1751. (2 vols)
The empirical and handicraft environment from which the science of chemistry was gradually emerging is reflected in this book. The author tells us that his previously published Elémens de chymie theorique was for persons "who have no knowledge of chemistry." Practique, on the other hand, is concerned with "the fundamental truths enunciated in Théorétique." But the first chapters deal with eminently practical topics, vitriol, alum, sulphuric acid and its concentration. And yet the system contemplated finds no place for such topics as saltpeter, potash and sal ammoniac. The second volume is devoted to animal and vegetable chemistry.

8. DIDEROT, Denys & J. le D. D'ALEMBERT, eds. Encyclopédie. Paris, 1751-1765. (17 vols)
This celebrated work (particularly for its illustrations, so far as technology is concerned) deals with the conventional industrial chemicals. It may be on a higher scientific level than the encyclopaedias of Zedler and Krünitz (nos. 4 & 16), but is far less detailed. The article on saltpeter, for example, is limited to 8 pages.

9. JUSTI, J.H.C. Abhandlung von den Manufakturen und Fabriken. Berlin, 1757-1761. (2 vols. New edition by Johann Beckman, Berlin, 1780)
A section on general considerations is followed by another on manufactures (concerned with textiles) and on factories. The latter is subdivided into gold and silver, iron and steel, other metals, porcelain and glass, and "mineral salts and colors." The mineral salts are alum, vitriol, sulphur, arsenic, and mercury compounds. The colors are smalt, cinnabar, white lead and verdegris. There is finally a miscellany dealing with leather, sugar, tobacco and beeswax. I have used the 1780 edition. Justi also began the Schauplatz der Künste and Handwerke, Königsberg, Nürnberg, Berlin, 1762-95 (17 to 20 vols), a translation of the French Descriptions (no. 13).

10. HOFFMANN, G.A. Die Chemie zum Gebrauch des Haus- Land- und Stadtwirtschaften, der Kunsten, Manufacturien, Fabricanten und Handwerken. Gotha, 1758 (and in the journal Oeconomische Nachrichten, 10, (1758), 1-286).
Intended as a schoolbook for apprentices in chemical trades outside the established metallurgy and pharmacy, the book deals succinctly with the full range of such industries. The author says that the arrangement by substances was decided upon after considering alphabetical arrangement or arrangement by trades. The book won a prize offered in 1757 by Peter Freyherr von Hohentz for a text on this subject, and enlarged editions subsequently appeared, under different titles and sometimes even under different authors. The well-known J.C. Wiegleb brought out a version called Anleitung zur Chemie für Künstler and Fabrikanten, of which I have seen the second edition (1779). Hoffman was a lawyer.

11. "Versuch eines chymischen Lehrbegriffs zum Gebrauch des Oecomomen." Oeconomische Nachrichten, 12, (1759-60) 107-94, 207-41, 341-54, 401-27; 13, (1761) 363-434, and Neue Oeconomische Nachrichten, 1, (1763) 429-560.

    A miscellany of helpful information (e.g., how to tell when wine fermentation is complete), a discussion of the chemical aspects of agriculture and of cooking, with long special treatises on milk and the brewing of beer. Longer (352 pp.) than Hoffmann's book (with which one suspects it competed for the above mentioned prize) it is also disorganized; but not without interest. The author bases himself on "facts," having no time for theory or experiment. He got his facts from reading and apologizes for his meagre chemical knowledge.

12. HALLE, J.S. Werkstädte der heutigen Kunst. Brandenburg & Leipzig, 1761-79. (6 Volumes. 2423 pp.)

    Treats most of the chemical topics, along with the rest of technology, although he omits some basic industrial chemicals (vitriol, alum, borax). His model was Tomasso Garzoni's history of trades (Piazza Universale, 1586), a work of little chemical significance. In the preface to vol. 2 he notes that Justi has begun the translation of the similar French work, Descriptions des Arts et Métiers (no. 13), and laments being in competition with a whole academy. Nevertheless he persevered to the end of this work and in 1782 published an "expanded abstract" of the Werkstädte, under the title, Technologie, oder die mechanischen Künste (Brandenberg, 760 pp.). Chemistry was still a "mechanical art" to Halle, and he devotes about 40 pages here to the conventional chemicals, including most of those he missed in the Werkstädte. Halle was Professor to the Cadet Corps in Berlin.

13. DESCRIPTIONS DES ARTS ET METIERS. Paris, 1761-88.

    The idea of publishing descriptions of "trades," of which Francis Bacon was the most notorious advocate, was reflected in England is a small miscellany of books and articles published sporadically in various places. But in France it was pursued doggedly for over two decades under the above title, which covers treatises ranging in length from articles to several volumes. The project was never terminated; it petered

out. The result is an uncertainty as to how many
volumes were published. One of the more complete
sets, in the Smithsonian Institution, is bound in 45
volumes. The original was a lavish and expensive
publication, and a more modest reprint (or version)
was published in Neuchatel, 1771-83, in 19 fat volumes. The German translation, started by Justi, comprised 20 volumes. There was also a Dutch translation, by P.J. Kasteleijn. Books or articles in this
series cover, usually at substantial length, glues,
charcoal, wax, ceramics, soap, bricks, starch,
leather, paper and parchment, paints, dyes, cooking
and coal. One volume, by J.F. Demachy, is entirely
concerned with the chemical arts and is listed
separately below (no. 18). A guide to this collection
has been published by A.H. Cole & G.B. Watts, The
handicrafts of France, Cambridge, Mass.: Baker Library
of Harvard University, 1952.

14. LEWIS, William. Commercium philosophico-technicum, or
the philosophical commerce of the arts. London,
1763. (646 pp)
"Experimental inquiries" by a well-known London
chemist, on gold, glass and porcelain, water-powered
furnace blowers, colors and platinum, with a miscellaneous appendix and numerous historical asides.

15. MACQUER, P.J. Dictionnaire de chimie. Paris, 1766.
(2 vols. There are numerous later editions and
translations, including an English version of 1771)
In this famous book Macquer has progressed substan-
tially since his Elémens (no. 7) in disentangling the
science of chemistry from technology. But he admits
that utility "is the greatest advantage" of chemistry,
and advertises his inclusion of such useful topics as
coal, inks, bread, wine, beer, vinegar, enamel, pottery, dyeing and others. The best edition is the
German translation, by J.G. Leonhardi (Chemisches
Wörterbuch, Leipzig, 1788-90), with profuse notes
which have expanded the work to seven volumes. The
Germans here used a French book as a base for the
display of their immense learning, as they did in
other cases (e.g., Demachy [no. 18]).

16. KRUENITZ, J.G. Oeconomische Encyklopädie. Berlin,
1773-1858 (242 vols.)

Even though more restricted in scope than other encyclopaedias, this is the largest ever published, by a wide margin. Krünitz, a physician who preferred writing, had completed volume 73 (to Lei) at the time of his death. Those who carried on seem to have maintained his standards of completeness. The major chemical products are included. Saltpeter gets 295 pages in volume 131 (1822). "Salt" gets all of volume 134 and parts of those before and after it!

17. SPRENGEL, P.N. Kunst und Handwerk in Tabellen, Berlin, 1773-95. (17 vols)
Undaunted by the similar works in progess by Halle and Justi, Sprengel (whom I have been unable to identify) produced yet another massive history of trades. The usual chemical topics are included (saltpeter gets 54 pages), often with illustrations.

18. DEMACHY, J.F. Le distillateur liquorist et distillateur des eaux fortes. Paris, 1776. (153 pp.)
This is the most "chemical" volume in the Descriptions des arts et métiers (no. 13) and is possibly the most important book on industrial chemistry published up to that time. It deals with distillation products, organic (e.g., alcoholic beverages) and inorganic (e.g., the mineral acids), and also with "solid" products, such as salts. The products discussed here were to be the core of the forthcoming, more narrowly defined, chemical industry. Notes were added by (Heinrich ?) Struve to the Neuchatel edition of the Descriptions, and these were included, and augmented by J.C. Wiegleb and Samuel Hahnemann, in a German translation, Laborant im Grossen (Leipzig, 1784), which is altogether the most valuable edition.

19. BECKMANN, Johann. Anleitung zur Technologie, oder zur Kentniss der Handwerke, Fabriken und Manufakturen. Göttingben, 1777. (460 pp.)
Deals concisely with 324 Handwerke (crafts), Fabrike (Handwerke employing fire and the hammer on a large scale), and Manufakturen (other large-scale Handwerke), including most of the usual chemicals of commerce. This book popularized the use of the word "technology" in the sense which has prevailed ever since.

20. PFEIFFER, J.F. von. Lehrbegriff sämtlicher öconomischer und Cameralwissenschaften. Mannheim, 1777-79. (4 vols)
A general cameralist work with brief, but not insignificant, sections on the usual chemical manufactures, with particular reference to German practice. There is more of the same in the anonymous Die Manufacturen und Fabrike Deutschlands (Frankfurt/M, 1780, 2 vols) which is probably also by Pfeiffer.

21. WIEGLEB, J.C. Handbuch der allgemeinen Chemie. Berlin & Stettin, 1781 (2 vols., also later editions, including an "enlarged" English translation by C.R. Hopson, published as A general system of chemistry, theoretical and practical, London, 1789)
Deals comprehensively with pharmaceutical and technical chemistry, the latter including the following peculiarly named subjects, halurgy (salts), lithurgy (stones), hyalurgy (glass), metallurgic chemistry, zymotochemistry (fermentation industries), and phlogury (inflammable bodies). It concludes with "economical," meaning agricultural chemistry. Wiegleb was probably the most competent chemist to deal comprehensively with industrial chemistry up to his time.

22. WATSON, Richard. Chemical Essays. London, 1781-87 (and later editions, ranging from two to five volumes)
Contains substantial chapters on vitriol, saltpeter, salt, gypsum, lime, coal, and red and white lead. Especially informative on English practice. Watson was Professor of Chemistry at Trinity College, Cambridge, "although," according to a recent biographer, "he was completely ignorant of the subject."

23. SUCKOW, G.A. Anfangsgründe der ökonomische und technische Chemie. Leipzig, 1784 (717 pp.)
Four hundred pages of this large book are devoted to "general chemistry." Substances are treated briefly (saltpeter three pages) but intelligently and with footnotes; and the work is noteworthy for its systematic arrangement.

24. JUNG-STILLING, J.H. Versuch eines Lehrbuchs der Fabrikwissenschaft. Nürnberg, 1785. (636 pp.)
The context of this book concerns a debate in Germany

over the utility of factories. The author divides technology into chemical and mechanical and sees the peculiar need for factories in the fact that raw materials are raw. Hence his particular interest is in industrial chemistry. He is also said to be the first to subdivide the "chemical arts" into "wet and dry," the latter involving heat, as in metallurgy, lime-burning and pottery, the former the manipulation of liquids, as in the production of distillates, salts and acids. He deals at respectable length (saltpeter gets 29 pages) with most of the conventional chemical products.

25. ENCYCLOPEDIE METHODIQUE. Chemie, Pharmacie et Metallurgie. Paris. Vol. 1, 1786, Vol. 2, 1792.
This successor to the Encyclopédie of Diderot and d'Alembert, had the same publisher and covered most of the same topics. The volumes were however organized topically (the whole work included 199 volumes, published between 1782 and 1832, during which time the volumes on Chemie, etc. reportedly expanded to six). The two volumes dealt with here profited from a distinguished authorship of some articles, which were written by A.F. Fourcroy and L.B. Guyton de Morveau.

26. GMELIN, J.F. Grundsätze der technischen Chemie. Halle, 1786, (750 pp.)
This is probably the first complete guide with copious source citation to the full range of commercially important chemicals, animal, vegetable and mineral. In a new edition, called Handbuch der technischen Chemie (Halle, 1795-96, 2 vols) it approached 2,000 pages and was both detailed and up to date (saltpeter gets 59 pages). At the same time he published Chemische Grundsätze der Gewerbkunde (Hannover, 1795) in which he expanded yet further on these matters. Gmelin, the fifth of his famous family to be associated with the University of Tübingen, also wrote the first large history of chemistry.

27. KASTELEIJN, P.J. Beschouwende en werkende pharmaceutische, oeconomische en naturkundige Chemie. Amsterdam, 1786-94. (3 vols)
Volume three deals with "economical chemistry," subdivided into the peculiarly titled branches which had been used by Wiegleb (no. 21). The usual chemical

products are discussed, but there is unfortunately little emphasis on the Dutch chemical industry, which had probably been the most sophisticated in Europe in the 17th century. This is also true of the three-volume Chemische en physische oefeningen, which he published in Leyden, 1793-1797. In Kasteleijn's time the Dutch looked to France and Germany for information on chemistry.

28. LAMPRECHT, G.F. Von. Lehrbuch der Technologie. Halle, 1787.
This book, organized similarly to that of Jung-Stilling (no. 24) is also particularly preoccupied with chemistry, to which about three-fourths of its pages are devoted. His basic subdivisions are into products of crystallization (salts), fermentation (starch, wine etc.), distillation (brandy, nitric acid) "cooking," (soap, dying), bleaching, burning (lime, ceramics) and smelting (glass, metallurgy). It is especially notable for its voluminous bibliographies.

29. KUNRADT, J.G. & Joh. BEKMANN (sic). Anleitung zum Studium der Technologie, oder kurze...Beschreibung verschieden Kunste und Handwerke. Brünn, 1789.
Another collection of trades, covering 52, including lime-burning, salt-making, gunpowder, potash and saltpeter. They say that they "pass over" mineral alkali, without saying why.

30. ROESSIG, C.G. Lehrbuch der Technologie. Jena, 1790.
In this book, chemistry and mechanics are made part of a common category of Manufakture and Gewerke (the latter a term for the crafts, derived from mining). The former is subdivided into hand and machine work, paper and leather being distributed between the two. Gewerke are chemical by the wet way (bleaching, fermentation, salts, sugar, etc.), chemical by the dry way (lime, sulpher, metallurgy), mechanical (mills), and Fabriken. Ceramics and colors are included under the latter. The descriptions are brief but informative. Roessig was a prolific cameralist writer.

31. NICHOLSON, William. A dictionary of practical and theoretical chemistry. London, 1795, 2 vols., 2nd edition, 1808).
The "practical" part of this book is negligible. It

is included here primarily to illustrate the trivial character of British literature on chemical literature at this time, as compared to the French, and specially to the German. And yet the actual industry in Britain was the envy of the French and Germans!

32. TROMMSDORFF, J.B. Systematisches Handbuch der gesamt Chemie. Erfurt, 1800-07. (8 vols)
Trommsdorff directed a "chemical, physical and pharmaceutical boarding school" in Erfurt which is said to have trained at least 32 men who subsequently established chemical industries. Trommsdorff himself established a chemical factory in 1813. Notwithstanding this, and despite the absence of "chemical philosophy" from the book (he says there wasn't any), the discussion of individual substances is brief and of little interest. Most noteworthy is a "history of Galvanism" contained in volume five, surely one of the earliest histories of this topic.

33. FISCHER, Wilhelm. Chemische Grundasätze der Gewerbs-Kunde, oder Handbuch der Chemie für Fabrikanten... Berlin, 1802.
Fischer was one of Hermbstadt's students, but whereas his master doubted the existence of "chemical philosophy," the student doubted the possibility of teaching the empirical aspects of the trade. So this book deals primarily with the theory behind the trades but not very successfully. He subdivides the chemical trades after the model of Wiegleb.

34. RHEES, Abraham. The New Cyclopedia. London, 1802-20 (39 vols. There are other editions).
With this work the English reader possessed an encyclopedia which dealt extensively with science and technology, such topics as saltpeter (8 columns) being supplemented with articles on the several nitrates. It is larger and more authoritative than the Encyclopedia Britannica, the 4th edition of which appeared in 1810 in 20 volumes. Chemistry in the Britannica was chiefly described by James Tytler, a "scientific dabbler" who wrote three-fourths of the second edition (10 vols., 1778-89). Rhees, although not obviously better versed in chemistry, was a professional encyclopedist, having begun this career in revision of Chambers Encyclopaedia, the oldest of the English encyclopaedias.

35. WEISE, J.C.G. Oekonomische Technologie, oder vollständige Anweisung zur Anlegung und Betreibung der jenigen Gewerbe, welche mit der Landwirtschaft verbunden werden können. Erfurt, 1803. (2 vols)
By "economical technology," he signified, as did Wiegleb (no. 21), the chemical products of agriculture. These are sugar, saltpeter (and gunpowder, 44 pages), potash, starch and beer, all of which are treated extensively. Weise himself appears to have practiced some of these semi-domestic industries.

36. GOTTHARD, J.C. Vollständiges Handbuch der practischen Technologie. Hamburg, 1803-05. (2 Vols)
Modeled after the work of Jung-Stilling (no. 24) but modified, according to the author, by the recent tendency to subdivide technology into hand or machine operations and wet or dry chemical operations. The mechanical section seems never to have appeared, but the chemical section appeared not only in this form but as separate books, entitled Alaun und Vitriol-Siederei, Die Salpeter und Pottaschensiederei, Die Salzsiederei, and Die Kunst des Bierbrauens. Other topics are included in one or another of these, notably woad (a blue vegetable dye), vinegar, tobacco, brandy, nitric acid, and sugar. Saltpeter gets 49 pages. Gotthard, a professor of cameralist subjects at Erfurt, betrays an almost maniacal bent for classification, although it cannot be judged successful on the basis of the organization of this book.

37. HILDEBRANDT, Friedrich. Encyklopädie der gesammten Chemie. Erlangen, 1803-06. (1540 pages)
Issued in installments, the parts of this book seem rarely to have been collected without lacunae. About one-third is on technology, and that organized, after Wiegleb, according the Halurgie, Phlogurgie, etc. It is useful principally in giving some attention to new industries, Aërurgie (the production of gases), Hydrurgie (water, pure and mineral), and the nascent artificial soda industry.

38. KRUENITZ, G.F. Handbuch von Manufaktur- Fabriken- und Handwerks-Sachen. Berlin, 1804.
This book is noteworthy for the inclusion of extensive data on the prices of chemicals. Otherwise it is remarkable only for its peculiarities. After conventional injunctions on the proper classifications of

chemical and mechanical operations, he drifts, while discussing the price of wood on p. 35, into a general account of prices--from astrolabes to dog houses--and this occupies most of the remainder of its 252 pages! Krünitz (not to be confused with J.G. Krünitz [no. 16]) was a government lawyer in Berlin. He recommends the book as a Christmas present.

39. ROESSLING, C.L. Neue Fabriken-Schule. Erlangen, 1808-08.
Three parts, which are essentially separate books, Ueber Pottaschen und Salpeter Siedery (414 pp.), Ueber Stanniollägeren (tinfoil) und Hammerwerke mit Schwanzhämmern (432 pp.), and Die Fabrikation des Salmiaks und der dabey als Nebenprodukte gewinnbaren Fabrikante (694 pp). Graced with elegant, colored plates, these books provide full instructions for the establishment of large factories--just at the time the processes, and some of the products, were becoming obsolete.

40. POPPE, J.H.M. Handbuch der Technologie. Frankfurt a/M 1806-10. (4 Theile, 1058 pages)
Poppe, a student and disciple of Johann Beckmann, took up where the master left off and was even more prolific. In this book he follows the now conventional German division into mechanical and chemical (wet and dry) operations but modifies it in adding "mechanical-chemical" operations as brewing, dyeing, sugar, salt, pottery and lime-burning. The entries are brief but informative. The same may be said of his Technologisches Lexikon (Stuttgart & Tübingen, 1816-20. 5 Vols.), Technologisches Universal Handbuch (Leipzig, 1837), Ausführliche Volks-Gewerblehre (2nd. ed., Stuttgart, 1836--a textbook resembling that of Jacob Bigelow [no. 50]). On his historical writing see no. 136.

41. CHAPTAL, J.A. La Chimie Appliquée aux Arts et Manufactures. Paris, 1807. (4 vols. English translation, 1807).
This is the first book to reflect fully the impact of the new science of chemistry on the chemical industries. It is essentially a scientific book, and organized accordingly. But Chaptal, a chemist, government official, and later chemical manufacturer, was as practical as he was learned. He subsequently

published on French industry, especially its apparent lag behind that of Britain (see nos. 208, 211).

42. HINTERLANG, Carl von. Technologie oder Gewerbkunde. Munich, 1810.
The preface to this curious book identifies it as a history of the rise or origin of "240 Künstlern, Fabrikanten und Handwerken." They are subdivided into those relating to the animal kingdom, to the plant kingdom and to the mineral kingdom, and each of these is further subdivided into Material-Kunde and Waren-Kunde, which appears to refer to raw and finished products. Thus potash and gypsum appear as Material-Kunde, the famous dye, Prussian blue, as Waren-Kunde. Most of the common chemical products are described briefly but informatively.

43. THENARD, L.J. Traité de Chimie. Paris, 1813-16. (4 Vols., also later editions).
Although ostensibly a textbook of chemistry, this work is as biased towards the production of commercially important materials as was Chaptal's La Chimie Appliquée (no. 41) of a few years earlier.

44. HERMBSTADT, S.F. Grundries der Technologie. Berlin, 1814. (781 pp.)
Comprehensive, brief, and important coverage, especially of textiles and chemicals, including such unusual arts as parchment-making, chagrin, and glue. Hermbstadt, Professor of Technology at the University of Berlin when it was founded (1810), was personally interested in chemical factories.

45. PARKES, Samuel. Chemical Essays. London, 1815, (5 Vols)
"Principally relating to the arts and manufactures of the British dominions." As the title suggests, this is not a systematic treatise but a collection of articles of the author's choosing. But where he chooses to deal with a subject he does so at length, with much anecdote and a felicitious and chatty style unusual in chemistry. His chemical topics, all emphasizing commercial and industrial considerations, are calico printing, barytes (barium sulphate, used in textiles and coloring), carbon (charcoal and coal), sulphuric acid, citric acid, soda and potash, pottery and porcelain, glass, bleaching, (pure) water, and sal

ammoniac.

46. LAMPADIUS, W.A. <u>Grundriss der technischen Chemie</u>.
    Freyberg, 1815. (465 pp.)
    The system of classification initiated by Wiegleb (Halurgie, Lithurgie, etc.) is here rationalized and extended, with the inclusion of plant, animal and "atmospheric" chemistry, and the separation of gunpower, bleaching, dry cleaning (Fleckenausmachen), dyeing, leather, fermentation industries, soap, paper, and cements and glues into a class of industries involving "several bodies in combination." Brief but informative; one of the best books of its kind. Lampadius was Professor of Chemistry and metallurgy at Freyberg (i.e., Freiberg, Saxony).

47. <u>DICTIONNAIRE TECHNOLOGIQUE</u>. Paris, 1822-25. (22 Vols., text. 2 Vols., plates)
    Published during the transition from empirical to "scientific" industry, this large work is especially valuable. Saltpeter, for example, is now treated among the nitrates, but the 26 pages alloted to it recapitulate much of the empirical history of the previous century. Two of the five editors specialized in chemistry, P.J. Robiquet and Anselm Payen, both teachers in Paris, members of the "Institute" which succeeded the Académie des Sciences, and both themselves involved in chemical manufactures.

48. GRAY, S.F. <u>The Operative Chemist</u>. London, 1828. (881 pp.)
    He found it "astonishing" how little chemical practice in his time differed from "the old practice," and lamented the dependence of "practical chemists" on theoretical treatises or on digging the subject out of "a heap of extraneous matter" in dictionaries or encyclopedias. Here he has dug it out for us, mostly from "the two great French encyclopedias," the <u>Dictionnaire Technologique</u>, (no. 47) "now in progress," and another unnamed, probably the <u>Encyclopedie Méthodique</u> (no. 25). This is a kind of recipe book, covering everything but biased towards laboratory operations rather than commerical chemicals. Saltpeter gets only two pages but gunpowder and fireworks get 17. The book appeared in the United States in 1830, in the form of <u>The Chemistry of the Arts</u> by A.L. Porter. This was only "on the basis" of Gray's book, but has the same

number of pages and changes too subtle to be immediately apparent.

49. DUMAS, J.B.A. Traité de Chimie appliquée aux arts. Paris, 1828-45. (8 vols)
It is perhaps appropriate, in view of the status of the author, that science tends to overwhelm technology in this volume, which treats of nearly every substance known, with occasional references to use. Tacked on here and there, however, are genuine articles on chemical technology. These include combustibles (coal, charcoal and coke), gas and oil lighting, glass, pottery, saltpeter and gunpowder (85 pages), metallurgy, sugar, bread, beer and liquors, soap, tanning and dyeing (to which he devotes 286 pages). There is a German translation, but the time when the Germans displayed their erudition by embellishing French books had passed.

50. BIGELOW, Jacob. Elements of Technology. Boston, 1829. (520 pp., 2nd. ed., 1831)
Lectures delivered at Harvard University "on the application of the sciences to the useful arts." Chemical topics include, "the materials used in the arts...arts of illumination...dividing and uniting solid bodies...vitrification...induration by heat (ceramics)...preservation of organic substances...and metallurgy." A few earlier works are cited, but this is, as the topics suggest, a work of considerable originality.

51. GUILLOUD, G.L. Traité de Chimie Appliquée aux Arts et Métiers, et Principalement à la Fabrication des Acides Sulfurique... Paris, 1830. (2 vols)
Of some use as an indicator of contemporary practice, but not very remarkable, even on sulphuric acid. I have been unable to identify the author.

52. SCHUBARTH, E.L. Elemente der technischen Chemie. Berlin, 1831
I have seen only the second edition, which is called Handbuch der technischen Chemie (1839-40, 3 Vols.). Useful for contemporary practices. The author was a professor at the Gewerb-Institut and sometime government official in Berlin.

53. URE, Andrew. A Dictionary of Arts, Manufactures, and Mines. London, 1839 (and later editions)
Ure, an M.D., was primarily occupied with chemistry. He published a Dictionary of Chemistry, an updating of Nicholson's, in 1831, which immediately appeared in an American verison, as did the Dictionary of Arts. Neither, however, is particularly useful (saltpeter gets only three pages). Almost immediately, however, Ure added a book length appendix on "recent improvements," which is of some value. Ure gained fame as an apostle of industrialism.

54. PARNELL, E.A. Applied Chemistry. London, 1844. (2 vols)
Contains long sections of gas illumination, wood preservation, dyeing and calico printing, glass, starch, tanning, rubber, borax, soap, sulphur and sulphuric acid, and soda. Parnell was not highly regarded by his chemical contemporaries, but this rather old-fashioned book contains much historical information.

55. MUSPRATT, J.S. Chemistry, Theoretical, Practical, and Analytical. Glasgow, n.d. (1845-60). (2 Vols)
Strong on topics like glass, soap, sugar, etc; weaker on the basic chemicals, but an attractive and informative book. In fact its elegance suggests that it was intended for the library rather than the laboratory. The author was a British student of Liebig.

56. KNAPP, F.L. Lehrbuch der chemische Technologie. Braunschweig, 1847-48. (2 vols. and later editions; English translation, 1847)
Covers the conventional topics (with the exception of saltpeter), "in the form of groups of monographs upon particular branches of the chemical industry," according to the British translators. The result is not altogether rational, sulphuric acid and sulphur appear under "alkalis," and Nordhausen sulphuric acid is under vitriol. Still this is the best book on the subject to appear to this time. The British editors expanded the section on "combustibles" to add material on coal, and the American version adds more material

on "American coal." Guncotton (cellulose nitrate), a topic "only two years old," is included. Knapp was Professor of Technology at Giessen, then of Chemical Technology (the first to hold this title?) at Munich, where he also directed the Royal porcelain manufactory.

57. PAYEN, Anselme. <u>Précis de Chimie Industrielle</u>. Paris, 1849. (622 pp.) (and four later editions to 1859)

He regarded the recent multiplication of the applications of chemistry as a compliment to theory, and organized the book accordingly, including material (e.g., on the halogens) which was then more potentially than actually of commercial importance. Nitrogen, with virtually no uses, is included, but saltpeter is not. A third of the book is devoted to organic materials. The information in this textbook is considerably less than in the <u>Dictionnaire Technologique</u> (no. 47) in which Payen had also been involved.

58. WAGNER, J.R. <u>Die chemische Fabrikindustrie</u>. Leipzig, 1869. (2 vols.)

This book is rich in historical information on most of the important commercial chemicals and was the last textbook of which this could be said. It is a curiously old-fashioned sequel to several distinctly modern productions by Wagner himself. These included a <u>Hand- und Lehrbuch der Technologie</u> (1858-64) in which he dealt with improvements in the chemical industry, among others, as revealed in the "great exhibitions." In 1855 he had founded the influential abstracting periodical, <u>Jahres-Bericht über chemischen Technologie</u>, and in 1850 he had published a <u>Handbuch der chemischen Technologie</u>, which deserves to be called the first modern textbook on the subject. Before the end of the century it had reached a 13th edition and had been translated into English by William Crookes. It contains virtually no historical information.

From this time textbooks of chemical technology proliferated, and the authors gave increasingly less attention to "obsolete" practices. Showing little awareness that the industry had a history, they began the practice of "updating" and revision, which led in a straight line to the textbooks and encyclopedias of today. The straight line is not, however, obvious from titles. The English translators of Knapp's Lehrbuch (no. 56), made it the basis for the rather chaotic Chemical technology of Edmund Ronalds, Thos. Richardson and H. Watts (1855-67), a multi-volume text which describes itself as one volume in five parts. This was in turn "incorporated" into C.E. Groves & Wm. Thorpe, Chemical technology (1899-1903. 4 vols), the rationale of which is suggested by the inclusion of electric lighting, possibly because gas lighting had been one of the favorite subjects of such books. While devoid of history and not very strong on science these books are mines of curious incidental information.

Wagner, rather than Knapp, was the father of the modern textbook in Germany, through his Handbuch der chemischen Technologie of 1850. It had a life of its own, for new editions continued to appear after Wagner's death in 1880. On the other hand, Muspratt's Chemistry (no. 55) was translated into German and "improved" beyond all recognition by F. Stohmann and F. Kerl, into an Encyklopädisches Handbuch der technischen Chemie, which appeared in 10 large volumes at Braunschweig, 1888-1905. It gave occasional birth to supplements until supplanted by the Enzyklopädie der technischen Chemie of Fritz Ullmann, first in 12 volumes, 1914-22, in a second edition, 10 volumes, 1923-32, and recently in a third edition. The volumetric championship, however, appears to have passed to the Americans, to R.E. Kirk and D.F. Othmer, whose Encyclopedia of Chemical Technology comprised 15 volumes published between 1947 and 1956, and 22 volumes in a second edition which appeared 1963-72. Such is the degree of completeness of the Ullmann and Kirk-Othmer encyclopedias that some history, particularly on the multitude of substances of very recent commercial application, has crept in.

The journal of industrial chemistry and the open-ended series of books on special topics dates from about the same time. The industrious J.R. Wagner pioneered the former, with his Jahres-Bericht über chemischen Technologie, a periodical, reviewing current work, which he largely wrote himself from 1855 to his death, after which it was carried on by others. In the early years, when the number of commercially important chemicals was relatively small, this work is of great historical importance, as a year-by-year account of what was going on. The open-ended series of books was essentially introduced by Pompjus Bolley, whose Handbuch der chemischen Technologie, unlike earlier "handbooks" was a series title for books on special topics by various authors. It continued to spawn new volumes intermittantly, to the despair of bibliographers ever since. Some of the books were very distinguished; the famous Theoretical and Practical Treatise on the Manufacture of Sulphuric Acid and Alkali of George Lunge (London, 1896, 3 Vols.) first appeared in German in this series. But they are almost devoid of historical references. Their value depends upon their minute descriptions of current practices.

REPORTS ON INTERNATIONAL EXHIBITORS

The comprehensive exhibition of technological innovation began at the Great Exhibition of London, 1851, and ended with the St. Louis exhibition of 1904, the last international exhibition of this kind. The voluminous and variegated publications stemming from these events included not only conventional catalogues, but essays, sometimes of great length, on the state of various fields. Since these were usually historically oriented, the following, dealing with industrial chemistry, are worth consulting.

LONDON, 1851

59. KNIGHT, Charles. <u>Cyclopaedia of the Industry of All Nations</u>. London, 1851. (1810 pp.)
   The author was a journalist, interested in everything. Most chemical industries receive a brief but up-to-date discussion, in alphabetical arrangement.

60. <u>Lectures on the Results of the Great Exhibition of 1851</u>. London, 1852.
   Two lectures on industrial chemistry, by Jacob Bell and Lyon Playfair, are included in vol. 1: 133-99.

61. ZOLLVEREIN <u>Amtlicher Bericht über die Industrie-Austellung aller Völker zu London</u>, 1851. Berlin, 1852. (3 vols)
   Chemistry and pharmacy are discussed in vol. 1: 262-92, which begins with the observation that the British chemical industry produces enormous quantities of only a few products, whereas the German industry is more varied.

62. FRANCE <u>Travaux de la Committée Française sur l'industrie des nations. Exposition Universelle de 1851</u>. Paris, 1858. (8 vols in 12)
   Perhaps the most voluminous of all "world's fair" publications, this deals at length with glass, leather, soap, etc., but does not appear to recognize any "chemical industry."

63. <u>Album de l'exposition universelle</u>. Paris, 1859. (2 vols).
   Arts chimiques are treated in 2: 275-353, including an interesting introduction.

LONDON, 1862

64. <u>Reports of the Juries; Class II, Sec. A</u>. London, 1863. (200 pp.)
   The major chemical industries are reported on by the celebrated chemist, A.W. von Hofmann, with attention both to historical background and to recent innovation.

65. ZOLLVEREIN <u>Amtlicher Bericht über die Industrie- und Kunst-Austellung zu London</u>, 1862. Berlin, 1865. (750 pp.)
Chemische Produkte are treated on pp. 663 ff.

PARIS, 1867

66. SMITH, J. Lawrence. <u>The Progress and Condition of Several Departments of Industrial Chemistry</u>. Washington, 1867, (146 pp.)
Separately published from the <u>Reprint</u> of the United States Commissioners to the Paris Universal Exhibition in 1867, Vol. 2. Smith declared that "no one can paint in too vivid colors the degree of indebtedness the civilized world is already under to the chemist, and no enthusiast can transcend in his wildest speculations what we are yet to realize."

67. GRANDEAU, M.L. "Arts chimiques," in <u>Revue de l'exposition de 1867</u>. Paris, 1867. (vol. 1: 707-1010)

68. HOFMANN, A.W. von. "Report on the Colouring Matters Derived from Coal Tar, Shown at the French Exhibition, 1867," in Max Riemann, <u>On Aniline and Its Derivatives</u>. New York, 1868, pp. 99-156.
Not seen.

69. CHEVALLIER, Michel, ed. <u>Rapports du jury internationale. Exposition Universelle de 1867</u>. Paris, 1868. (12 vols)
Chemical industries are the subject of vol. 7 (390 pp.)

70. K.K. OESTERREISCHE CENTRALE COMITE. <u>Bericht über die Welt-Austellung zu Paris in Jahre 1867</u>. Vienna, 1869. (6 vols)
Chemical industries are discussed in vol. 3: 231-520.

VIENNA, 1873

71. BAUER, A. "Die chemische Grossindustrie," in <u>Officieller Austellungs-Bericht. Weltausstellung 1873</u>. Vienna, 1874. Gp. III, sec 1. (39 pp.)

72. FRANCE, COMMISSION SUPERIEUR. <u>Exposition Universelle de Vienne en 1873</u>. Paris, 1875. (5 vols)
Industrial chemistry is treated in vol. 3:3-138, 325-72.

73. HOFMANN, A.W. Von, ed. <u>Bericht über die Entwicklung der chemischen Industrie während des letzen Jahrzehends</u>. Braunschweig, 1975. (2 vols)
Nearly 1,600 pages on the development of industrial chemistry over ten years! But it was a crucial decade, and the editor was one of the most important participants. Originally published as vol. 3, pt. 1 of the German <u>Bericht über die Wiener Weltausstellung im Jahre 1873</u>.

74. SMITH, J. Lawrence. "Chemicals," in <u>Reports, United States Commission to the Vienna Exhibition, 1873</u>. Washington, 1876, 2:5-8.
Very cursory.

PHILADELPHIA, 1876

75. VEREINIGTEN CHEMISCHEN FABRIKANTEN, <u>Die chemische Industrie Deutschlands, auf der Weltausstellung in Philadelphia, 1876</u>. Berlin, 1876. (72 pp.)
Says there were relatively few German exhibitors because "the protective tariffs of the United States unfortunately make impossible the export of a large part of the chemical products of Germany." German exhibitors were nevertheless numerous. Miscellaneous information.

76. GOLDSCHMIDT, Guido. "Die chemische Industrie," in <u>Bericht über die Weltaustellung in Philadelphia 1876</u>. Vienna, 1877. Vol. 1, Heft 7. (43 pp.)
Heft 8 begins with a 20-page history of the American petroleum industry.

77. MALLET, J.W. "Chemistry and pharmacy," in <u>Reports of the United States Centennial Commission, International Exhibition, 1876</u>, Washington, 1880. (11 vols) vol. 4: 2-283.

PARIS, 1878

78. GIRARD, Charles, "Les produits chimiques pour la grande industrie à l'Exposition Universelle de 1878," La Nature, 6 (1877), 177-79, 242-47.

79. JENKINS, Thos. E. "Chemical and pharmaceutical processes," in Reports of the United States Commissioners to the Paris Universal Exposition, 1878. Washington, 1880. Vol. 4: 1-162.

80. LAUTH, Charles. Rapport sur les produits chimiques et pharmaceutiques. Paris, 1881. (366 pp.)
The introduction concludes with an expression of "disquiet" over the "gigantic effort" evident in the German and Swiss chemical industries. This is a separate publication of Gp. 5, Class 47 of France, Ministère de l'Agriculture et du Commerce. Rapport... Exposition Universelle de 1878.

PARIS, 1889

81. PICARD, Alfred, ed. Rapport générale...Exposition Internationale de 1889. Paris, 1892. (9 vols)
Chemical products and industries are treated in vol. 6:180-287, 469-501.

CHICAGO, 1892

82. FRANCE. MINISTERE DU COMMERCE. Rapport...produits chimiques et pharmaceutiques...Exposition Internationale de Chicago, 1893. Paris, 1894. (216 pp.)
The introduction (by Albin Haller) reviews the history from 1877 to 1893, giving "moral, economic and scientific" reasons for German supremacy.

83. WITT, O.N. Die chemische Industrie auf der Columbische Weltaustellung zu Chicago und in den Vereinigten Staaten in Jahre 1893. Berlin, 1894. (148 pp.)

84. ASSOCIATION OF THE CHEMICAL FACTORIES OF GERMANY. Guide through the exhibition of the German chemical industry, Columbian Exposition in Chicago. Berlin, n.d. (109 pp.)
Useful for statistics, and especially as a guide through the labyrinth of German firms, mergers, etc.

PARIS, 1900

85. VEREIN ZUR WAHRUNG DER INTERESSEN DER CHEMISCHEN INDUSTRIE DEUTSCHLANDS. Weltausstellung zu Paris, 1900. Berlin, 1900. (215 pp.)
Begins with a short discussion of the state of the industry, followed by information on each firm exhibiting, with brief histories of most of them, a notable feature of this publication.

86. HABER, Fritz. "Die Elektrochemie auf der Pariser Weltausstellung," ZAC, 14 (1901), 184-92, 215-26.
In effect this is a capsule history of this newly applied technology. Citations.

87. Report of the Commissioner General for the United States to the International Exposition, Paris, 1900. Washington, 1901. (6 vols)
Vol. 3: 5-98 includes a perfunctory notice of the "Department of liberal studies and chemical industries." (!)

88. KEPPLER, Gustav, "Chemisches auf der Weltausstellung in Paris im Jahre 1900," SC, 6 (1901), 1-38.
A useful summary, with a few citations.

89. HALLER, Albin, "Industrie chimique," in (France) Ministère du Commerce etc. Exposition Universelle de 1900. Rapports du Jury Internationale. Paris, 1902. (Classes 87, 402 pp.)

ST. LOUIS, 1904

90. VIEWIG, Walter, "Die Chemie auf der Weltausstellung zu St. Louis, 1904," SC, 10 (1905), 147-242.
A comprehensive description, arranged by countries. The introduction remarks that something outstanding generally appears at these expositions, aniline dyes at London, 1862, the Solvay process at Paris, 1867, synthetic alizarine at Vienna, 1873, the petroleum industry at Chicago, 1893, synthetic indigo at Paris, 1900, and now the contact process for sulphuric acid.

91. COHN, Paul, ed. Die chemische Industrie...Weltaustellung St. Louis, 1904. Vienna, 1905. (112 pp.)
Although the exhibit space was four times that at Paris in 1900, he concludes that no "epochal" novelties were shown. Describes the chemical exhibits, country by country, calling the American chemical industry inferior, which he attributes to education. He also remarks that the American "bureaucracy" did not favor the learned (Lernender) with much information--and this was, in fact, the last "world fair" at which information was competitive with entertainment.

GENERAL HISTORICAL LITERATURE

92. ASTARITA, Gianni. "L'Evoluzione dei fondamenti teorici dell'ingegneria chimica," Quaderni dell' Ingegnere Chimico Italiano, (supplement to) Chimica e Industria, (1972), no. 8: 112-14.
Calls "prehistoric" the period before 1898, where he first encounters the name, chemical engineering. A second era began with the 20th century, being particularly noted for the recognition, in the United States about 1930, of the similarity of processes which were diverse in terms of their products. A third era began with the introduction of the concept of "transport phenomena" in the 1950's. No citations.

93. BADISCHE ANILIN UND SODAFABRIK. firm. <u>Schriftenreihe der Unternehmensarchivs der BASF.</u> Ludwigshafen, 1963 ff.
    Occasional (usually annual) publications based on the archives of this important firm. Those I have seen are entered separately (abbreviated <u>SB</u>).

94. BECKMANN, Johann. <u>Beyträge zur Geschichte der Erfindungen.</u> Leipzig, 1780-1805. (5 vols. There were English translations from 1797)
    Its charming and informative chapters range from a history of Italian bookkeeping to a history of the canary bird, but include some on industrial chemicals, namely alum, aqua fortis, indigo, saltpeter, sal ammoniac, soap, smalt and verdegris.

95. BERTHELOT, Marceliin. "L'Evolution générale des méthodes dans les industries chimiques," <u>Revue Général des Science</u>, 11 (1900), 869-75.
    An address to the International Congress of Applied Chemistry, in Paris, by the grand old man of French chemistry. No citations.

96. BINZ, Arthur. "Ueber den Ursprung der chemischen Grossindustrie," <u>ZAC</u>, 25 (1912), 2337-39.
    Sees its origin in the large-scale importation into Europe of American cotton, from 1791-93.

97. BIRD, R. Byron et al. "The Role of Transport Phenomena in Chemical Engineering Teaching and Research: Past, Present, and Future," in <u>Furter</u> (1980, no. 116) 153-66.
    "Transport phenomena...the transport of mass, momentum, energy, and other entities" is seen as a post-World War II phenomenon, with roots as far back as Newton. Describes the introduction of the concept at the University of Wisconsin. The numerous citations constitute something of a bibliography of the subject.

98. BROWNE, Charles A. "The chemical industries of the American aborigenes," Isis, 23 (1935), 406-24.
Discusses fertilizer, maple sugar, flour, chocolate, salt, tanning (with a mixture of cooked brains, liver, and grease), dyes, pottery, metallurgy, and medicines (curare). Bibliography.

99. BUCHNER, M. Die chemische Industrie und Metallurgie in den zweiter Hälfte unseres Jahrhundert. Graz, 1893.
Not seen.

100. BUCHOLZ, K. "Verfahrenstechnik (Chemical Engineering) --Its Development, Present State and Structures," Social Studies of Science, 9 (1979), 33-62.
An analysis of "the introduction of advanced procedures in chemical technology at the beginning of the present century," and its subsequent development in terms of university involvement, institutionalization, etc. Sees the institutionalization of Verfahrenstechnik in Germany as a post-World War II phenomenon. Citations.

101. CAMPBELL, W.A. The chemical industry. London, 1971, (156 pp.)
A sound introduction to its history although very brief on any particular topic and largely restricted to Britain--and extending only to about 1900. Citations.

102. CAPOCACCIA, A., ed. Storia della technica. Turin, 1973-80 (4 vols.)
Although it comes up to date, this encyclopedic work gives less than 100 pages to industrial chemistry, and a third of that is concerned with rockets and atomic energy.

103. CHRISTIANSEN, Carl C. Geschichte des Standorts der chemischen Industrie. Tübingen, 1913 (dissertation, 45 pp.)
Deals with factors determining the location and organization of chemical industries.

104. CLEMENTS, Richard. Modern Chemical Discoveries. London 1954. (200 pp.)
Strongly oriented towards chemical technology (the author was formerly editor of Chemical Age) but although up to date it is generally too brief to be useful on any particular topic.

105. CLOW, Archibald, & Nan CLOW. The Chemical Revolution: a Contribution to Social Technology. London, 1952. (680 pp.)
Although written in and from the point of view of Scotland this is one of the most comprehensive and circumstantial histories of industrial chemistry during its nascent period. Deals with the "heavy chemicals," salt, alum, alkali, and with traditional industries such as dyeing, glass, paper, sugar, soap, and beer. Citations.

106. COLSON, Albert. L'essor de la chimie appliquée. Paris, 1910. (349 pp.)
Contains some historical information but is primarily a popular account for the totally uninformed.

107. DAUMAS, Maurice, et al., eds. Histoire générale des techniques. Paris, 1962-79. (5 vols)
Those chapters devoted to chemical technology cover about 400 pages (nearly 60 percent of which is concerned with the 20th century). They differ from those in other recent encyclopedic histories in being written for the most part by professional historians of technology. Bibliographies. No citations.

108. DJERMANOVITCH, Rayko. Le Traité de Versailles et des matières premiere. Paris, 1927. (173 pp.)
On the role of raw materials in the First World War and its aftermath.

109. FERBER, Johann J. Neue Beyträge zur Mineralgeschichte verschiedener Länder. Mietau, 1778. (462 pp.)
Pt. 4, "Nachricht von einigen chymischen Fabriken," 317-90, describes the manufacture of vitriol (in England), verdegris (in Montpellier), and, mostly in

Holland, mineral acids, sulphur, sal ammoniac, borax, mercurial and lead products.

110. FERGUSON, John. <u>Some early treaties on technological chemistry</u>. Glasgow, 1888, (34 pp.)
Discusses a number of printed books, beginning with the early 16th century <u>Kunstbüchlein</u>. Most of this was published in the <u>Transactions</u> of the Royal Society of Glasgow from 1883 to 1885, and in its <u>Proceedings</u> for 1887-88. It supplemented his <u>Bibliographical notes on histories of invention and books of secrets</u>, Glasgow, 1883 and 1885, and was itself followed by several supplements to 1919.

111. FESTER, Gustav. <u>Die Entwicklung der chemischen Technik, bis zu den Anfängen der Grossindustrie</u>. Berlin, 1923. (225 pp.)
This modest book, the outgrowth of lectures at the University of Frankfurt a/M in 1919, is subtitled "Ein technologisch-historischer Versuch," because, as the author tells us, he had to accomplish in three years what should have required thirty. That it remains after nearly sixty more years the best book on the general history of chemical technology (in my opinion) tells us something about the state of the field.

112. FIGUIER, Louis. <u>Les merveilles de la science</u>, Paris, 1967-70 (4 vols), and <u>Les merveilles de 'industrie</u>, Paris, (1873-77), (4 vols)
These massive (5,844 pages) and elegantly illustrated books, written by a Paris journalist (presumably for the luxury trade) incorporate far more history than one would expect and much unusual information. Citation is sparse, but one receives a general impression of authenticity. Figuier did not trouble himself about differentiating "science" from "industry," and a given topic may occur under either title. Chemical subjects, often treated at great length, include alcohol (128 pp.), anesthetics, asphalt, beer, bleaching, bread, dyeing, food preservation, glass gunpowder, heating, lighting, leather, milk, oils, paper, phosphorus, photography, potash, pottery,

rubber, salt, soap, soda, sugar, sulphur, sulphuric acid, vinegar, water, and wine.

113. FLORENTIIS, G. de. Storia della technica. Milan, 1968. (2 vols)
Industrial chemistry is treated very briefly (56 pages), and mainly through biographies.

114. FORBES, Robert J. Studies in Ancient Technology. Leyden, 1955-64. (9 vols)
Chemical topics, usually treated extensively, and with citations, include cosmetics and perfumes, food, alcoholic beverages, vinegar, salt, natron, sal ammoniac, soap, and pigments (all in vol. 3); bleaching and dyes (vol. 4); leather, sugar and glass (vol. 5); and refrigerants (vol. 6).

115. FRESHWATER, D.C. "George E. Davis, Norman Swindin, and the Empirical Tradition in Chemical Engineering," In Furter (1980, no. 116) 97-113.
Davis (1850-1906), largely a "works chemist," is credited with the "unit operations" concept before 1888, as particularly embodied in his Handbook of chemical engineering (1901). Swindon (1880-1976, see his autobiography, no. 447) was his employee.

116. FURTER, Wm. F., ed. History of chemical engineering. Washington: American Chemical Society, 1980. (435 pp.)
Twenty-two articles, assembled on the "approximate centennial" of chemical engineering as "a distinct profession." Many are important and are entered separately.

117. GANZENMULLER, Wilhelm. Beiträge zur Geschichte der Technologie und der Alchemie. Weinheim, 1956.
Collects 17 previously published articles (1934-40) on chemical technology, nearly all on glass.

118. GUEDON, J.-C. "Conceptual and Institutional Obstacles to the Emergence of Unit Operations in Europe," in Furter (1980, no. 116), 45-75.
An important analysis of the history of the relationship between chemical education and the industry, in England, France, and Germany, during the 19th century. Citations.

119. GUINOT, Francois. Les stratégies de l'industrie chimique. Paris: Calmann-Lévy, 1975. (281 pp.)
Essentially not an historical work but significant (if not unique) in its attempt to analyze "the industrial revolution in chemistry 1965-73" and its causes. Citations.

120. HABER, Fritz. Fünf Vorträge aus den Jahren 1920-23. Berlin, 1924. (92 pp.)
Includes, "Ueber die Darstellung des Ammoniaks aus Stickstoff und Wasserstoff," "Die Chemie im Kriege," "Das Zeitalter der Chemie," "Neue Arbeitswessen," and "Zur Geschichte des Gaskrieges."

121. HABER, L.F. The Chemical Industry during the Nineteenth Century. Oxford, 1958. (292 pp.)
Technical and economic history, the best book yet published on the subject and period. Citations.

122. HABER, L.F. The Chemical Industry, 1900-1930. Oxford: Clarendon Press, 1971. (425 pp.)
A continuation of the previous work.

123. HAYNES, Williams. This chemical age. New York, 2nd ed., 1942. (401 pp.)
A popular description of the chemical industry with much historical anecdote and no citations. Largely restricted to the U.S.

124. HELBIG, E. Pharmacie und chemische Grossindustrie, ihre Entwicklung und volkswirtschaftliche Bedeutung. Tübingen, 1922. (diss. Tübingen Univ.)
Not seen.

125. HOECHST (firm). Documenta aus Höchster Archiv. Frankfurt/M, 1964 ff.
These admirable publications from the archives of this important company consist of small (ca. 50 pp.) monographs, reproduced from typescript on many topics. Some are here entered separately (abbreviated DHA).

126. IDHE, A.J. "Chemical Industry, 1790-1900," Journal of World History, 4 (1958), 957-83.
A brief but informative survey with references to secondary sources.

127. KOLB, J. Sur l'evolution actuelle de la grande industrie chimique. Lille, 1883.
Not seen. Published by the Société Industrielle du Nord de la France.

128. KRANZBERG, Melvin & Carroll PURSELL, Jr., eds. Technology in Western Civilizations, New York, 1967. (2 vols)
Gives 84 pages to industrial chemistry, in the 19th and 20th centuries by Eduard Farber and Robert Multhauf.

129. LIPPMAN, E.O. von. Abhandlungen und Vorträge zur Geschichte der Naturwissenschaften. Leipzig, 1906-13. (2 vols)

130. LIPPMAN, E.O. von. Beiträge zur Geschichte der Naturwissenschaften und der Technik. Berlin, 1923; Weinheim, 1955. (2 vols)
These four volumes reprint the majority of 254 articles on history published between 1878 and 1938 by one of the most prolific authors on the history of chemistry. The number on chemical technology is relatively small, and some have been entered separately below.

131. LITTLE, Arthur, D. "Chemical Industry," JCE, 5 (1938), 641-55.
A lecture, notable mainly for the source, a pioneer in the promotion of industrial science in the United States. No citations.

132. MACH, Erich. "Die Entwicklung in der Gestaltung chemischer Werke seit 1865, am Beispiel die BASF," Chemie-Ingenieur Technik, 35 (1963), 133-42.
An important discussion of changes in the style of construction, organization, and location of chemical works, especially as illustrated by the works of BASF from 1865 to the present. Illustrated. No citations.

133. MAIERHOFER, J. & A. DIEM, "The Origins and Subsequent Development of the Chemical Industry, Sandoz Bulletin, no. 1: 14-24; no. 2: 10-25; no. 3: 2-21 (all 1965), no 5: 12-25 (1966).
Subdivided into four parts, concerned with the soda industry, the mineral acids, dyestuffs and textile

auxilleries, and ammonia. Brief undocumented summaries, but authoritative and beautifully illustrated.

134. NATURHISTORISCHE UND CHEMISCH-TECHNISCHE NOTIZEN. Berlin, 1854 ff.
This annual publication, a cross between a news magazine and a recipe book, happily incorporated long sections on particular topics, combining some history with an evaluation of the state of the field. Among them were agricultural chemistry and fertilizers (4 [1856] 1-192), water glass (soluble silicates, 8 [1858] 1-43), dyeing (ibid., 58-355), sugar (10 [1859] 13-103), coal, illuminating gas and natural oils (14 [1861] 3-263) and photography (15 [1863] 23-172).

135. PARTINGTON, J.R. Origins and Development of Applied Chemistry. London, 1935. (597 pp)
A voluminous and immensely informative account of the chemical arts in antiquity, based primarily on archeological and paleographic evidence. Citations.

136. POPPE, J.H.M. Geschischte der Technologie seit der Wiederherstellung der Wissenschaften bis an das Ende des 18. Jahrhundert. Göttingen. (3 vols)
Poppe here follows his mentor, John Beckmann, with a more systematic work than Beckmann's History of Inventions (no. 94). The chemical topics included are paper, gunpowder, salt, sugar, bleaching, starch, dyes and dyeing, beer and brandy, ceramics and glass, most of them subjects of substantial articles. Poppe also published a Geschichte aller Erfindungen und Entdeckungen (Stuttgart, 1837) which is primarily concerned with personal and household articles and has little of chemical interest.

137. REILLY, Desmond. "Salts, Acids and Alkalis in the 19th Century," Isis, 42 (1951), 287-96.
A comparison of "advances" in French, German, and English industry; a whiggish account of the growth of "the chemical baby" into "a husky young adolescent."

138. SCHMAUDERER, Eberhard. "J.R. Glaubers Einfluss auf die Frühformen der chemischen Technik," Chemie Ingenieur Technik, 42 (1970), 687-96.
A summary remarkable for its ample documentation.

139. SCHMAUDERER, Eberhard. "Der Einfluss der Chemie auf
der Entwicklung des Patentwesens in der zweite
Hälfte des 19. Jahrhundert," Tradition, 16 (1971),
144-76.
Patent rights for German chemists were only firmly established in 1907, the first such action in behalf of any special group. Citations.

140. SCHMIDT, Albrecht. Die industrielle Chemie in ihrer Bedeutung im Weltbild. Berlin & Leipzig, 1934. (829 pp.)
Lectures given at the University of Frankfurt a/M in 1932 by one who had been active in the German chemical industry since the mid 1880's. Thus although it deals with the technical and economic state of the chemical industry in his time, his observations have a much stronger historical flavor than one finds in most books of this kind.

141. SINGER, Charles et al., eds. A History of Technology. Cambridge, 1953-78. (7 vols)
Industrial chemistry fares better in this encyclopedic work than in its contemporary rivals, receiving over a thousand pages. A larger variety of topics is also covered. Most of the authors were professional engineers rather than historians. Citations and numerous illustrations.

142. STRUBBE, Irene. "Industrielle Revolutionen und Chemie, ihre Wechselbeziehungen zwischen 1785 und 1860," NTM, 15 (1978), 39-47.
Sees the rise of a scientifically oriented chemical industry in the explosives and coal tar industries, at the end of the period covered--leading to "the end of the capitalism of free competition."

143. SWORYKIN, A.A. et al, eds. Geschichte der Technik. Leipzig, 1964. (831 pp., trans. from the Russian original of 1962)
Divides history into four periods, feudal, early and late capitalism, and "after the great October [1917] socialist revolution," the latter of which occupies about half the book. Chemical technology gets very short shrift in the first three periods and only 41 pp. in the last.

144. TAYLOR, F. Sherwood. <u>The History of Industrial Chemistry</u>. New York, 1957.
Mistitled. It is a rambling tour through pure and applied chemistry which does justice to neither. A poor book by an author who distinguished himself in many others.

145. TREUE, W. "Die Bedeutung der chemischen Wissenschaft für die chemische Industrie, 1770-1870." <u>Technikgeschichte</u>, 33, (1966), 25-51.
Concludes that the debt of industrial chemistry to science is such that we should speak of a scientific rather than an industrial revolution so far as chemical technology is concerned. Citations.

146. VERSHOFEN, Wilhelm. <u>Die Anfänge der chemische-pharmazeutischen Industrie, eine wirtschaftshistorische Studie</u>. Berlin, 1949. (3 vols, 426 pp.)
Emphasizes efforts at industrialization by J.B. Trommsdorff and E. Merck. Some references in text. An important but little-known work.

147. WILLIAMS, T.I. <u>The Chemical Industry, Past and Present</u>. London, 1953. (192 pp)
"Past" occupies pp. 11-82, interestingly and authoritatively although without references.

148. WITT, O.N. "Die Entwicklung der technischen Chemie," <u>Bericht DCG</u>, 40 (1907), 4644-52.
An upbeat but superficial recitation of the triumphs of chemistry during the previous 40 years. An English translation appeared in SIAR for 1908: 255-62.

149. WITT, O.N. "Wechselwirkungen zwischen der chemischen Forschung und der chemischen Technik," in Paul Rinneberg, ed., <u>Kulture der Gegenwart</u>, Th. 3. Abt. 3, Bd. 2, Leipzig, 1913: 475-527.
The relationship is "symbiotic:" a summary but excellent account of the 19th century chemical industry, aimed to demonstrate the importance of science. No citations. <u>Kultur der Gegenwart</u> may be the most massive and complicated of the giant literary celebrations of the accomplishments of "modern" Europe--published just in time.

150. ZART, A. Die Entwicklung der chemischen Grossindustrie. Munich, 1922. (38 pp.)
A cursory survey of recent developments. No citations.

151. ZIMMERMANN, P.A. Patentwesen in der Chemie - Ursprüng, Anfänge, Entwicklung. Ludwigshafen: BASF, 1965. (155 pp.)
A detailed analysis of the problems posed by chemical patents in the four decades prior to 1914. Citations.

152. ZIMMERMANN, P.A. Ueber die Grenzen hinaus. Notizen zur industriellen Entwicklung im 19. Jahrhundert. Ludwigshafen, 1971. (77 pp., SB no. 8)
An essay on the role of the European chemical industry in the internationalization of industry. Includes a chronology.

153. ZIMMERMANN, P.A. "Chemie-Politik-Fortschritt. Notizen zur Entwicklung einer Industriezweiges im Europa des 19. Jahrhunderts," Technikgeschicte, 41, (1974), 53-67.
Cursory but perspicaceous inquiry into the ascendency of Germany over Britain, especially in the coal tar industry.

ASIA

GENERAL

Because of the sparse and miscellaneous character of western literature (which alone is covered here) it has not seemed appropriate to subdivide the following into "early works," and "histories."

154. "Fabrication du bleu de Prusse en Chine," Annales de l'Industrie Nationale, 14 (1824), 105.
Says that the Chinese imported Prussian blue from Europe until 1819, when a factory was opened in Canton.

155. ATKINSON, R.W. "The Chemical Industries of Japan," Trans., Asiatic Society of Japan, 6 (1878), 277, 7 (1879) 313.
Deals with the manufacture of oshiroi (white lead) and ame (dextrose and maltose).

156. BARKER, Dee H., & C.R. MITRA. "A History of Chemical Technology and Chemical Engineering in India," in Furter (1980, no. 116, 227-48).
Very summary, since it covers 4000 years. But mostly concerned with the present state. The Indian Institute of Chemical Engineers was founded in 1947. Citations.

157. BLAIR, Dorothy. History of Glass in Japan. New York: Kodansha, 1973. (474 pp.)
A well-written narrative with many quotations, notes, and illustrations.

158. BRETSCHNEIDER, E. Medieval Researches from Eastern Asiatic Sources. London, n.d. (1887). (2 vols)
Discusses Chinese literary sources, including references to sal ammoniac from central Asian volcanos.

159. CHANG SIU-MING. "A Note on the Invention of Paper in China," Papiergeschichte, 9 (1959), 21-2.
Summarizes the state of a controversy. A few citations in the text.

160. CHEMICAL AGE OF INDIA. "Twenty Years of Indian Chemical Industry, 1949-1969," Chemical Age of India, 20 (1969), 985-1044. (whole of the December issue)
A chronology is followed by sections on each class of product. No references but valuable as a guide to the subject.

161. DE SOUSA, J.P. History of the Chemical Industry in India. Bombay, 1961. (307 pp.)
Largely concerned with economics and with the period since 1945, with a 45-page summary of history before 1939. No references.

162. DIVERS, E. "The Manufacture of Calomel in Japan," Journal, Society of Chemical Industry (London), 13 (1894), 108-11.
An eyewitness description of a production process of supposed antiquity, perhaps dating to the end of the first Christian millenium.

163. EYRE, John D. "Salt from the Sea: a Geographical Analysis of the National and International Pattern

of Japanese Salt Production and Trade." unpublished dissertation, Univ. of Michigan, 1951.
History is incidental but not insignificant; indeed it is the "old" Japanese industry which is described and which has subsequently virtually ceased to exist. Citations.

164. GALE, Esson M. Salt for the Dragon. East Lansing, Mich., 1953. (225 pp.)
An entertaining reminiscence of the Chinese salt industry, by one who arrived there in 1908 and was involved in foreign administration of the Chinese salt monopoly after the "Reorganization loan" of 1913 pledged the salt revenue of China as security for payment of the indemnities imposed on China after the "Boxer Rebellion."

165. GOIDSENHOVEN, J.P. van. La ceramique chinoise. Commentaires sur son évolution. Brussels, 1954. (213 pp.)
Historically organized, but principally in reference to art, with little on technology.

166. GOODRICH, L. Carrington. "Paper. A Note on its Origin," Isis, 42 (1951), 145, and "More on Paper," Ib., 44 (1953), 277.
He doubts claims for Chinese paper before the Christian era and mentions a fragment dated about 98 A.D.

167. HANBURY, Daniel. "Notes on Chinese materia medica," Pharmaceutical Journal, ser. 2, 2 (1860), 15-18, 109-16, 553-57, 3 (1861), 6-11, 204-09, 260-264, 315-18, 420-25.
Compares a printed Chinese pharmacopoeia with specimens of drugs from China which he subjects to a physical and and chemical analysis. Citations note European equivalents. In a postscript (ser. 2, 6 [1865] 514-15) he analyzes samples of Chinese sal ammoniac, which he thinks of volcanic origin.

168. JIMBO, Genji et al. "The History of Chemical Engineering in Japan," in Furter (1980, no. 116) 273-82.
Sketchy but informative. Kotaro Shimamura is known to have translated (but not published) G.E. Davis' Handbook of chemical engineering in 1903, two years after

its first publication in England. In 1940 there were 9 graduates in chemical engineering, from 3 institutions. In 1969 there were 1,204 from 35 institutions. Citations.

169. JULIEN, Stanislas. Industries ancienne et moderne de l'Empire Chinoise. Paris, 1869. (254 pp.)
Deals with coal, salt, lime, paper, foods, dyestuffs, sulphur, sugar and many other chemical materials, on the basis of notes from Chinese sources. Interesting contemporary illustrations.

170. JULIEN, Stanislas, translator. Histoire de la fabrication de la porcelain chinoise. Paris, 1856. (320 pp.)
Translation of a Chinese treatise on porcelain which had been published in 1637--and which has subsequently been attributed to P'u Lan. Includes a 74-pp. historical introduction by the translator (citations) and interesting illustrations of Chinese origin.

171. KOCHER, E. "Die Salzindustrie von Tse Liu Tsing," Technikgeschichte, 27 (1938), 116-22.
A description of the Szechuan salt industry by a German resident in China.

172. LAUFER, Berthold. The Beginnings of Porcelain in China. Chicago, 1917. (102 pp. Field Museum Publication 192, 79-181)
Dates porcelain from the late 6th century. Citations.

173. LI CH'IAO P'ING. The Chemical Arts of Old China. Easton, Pa., 1948. (215 pp.)
Comprehensive although chronologically vague. No citations, although the text frequently refers to Chinese writings by name. Illustrated.

174. LI JUNG. "Account of the Salt Industry at Tzu-Liu-ching chi," Isis, 39 (1948), 228-34.
Translation, by Lien-che Tu Fang, of a nineteenth century description (Li Jung died in 1889) of the salt industry of Szechuan province.

175. MAYERS, W.F. "On the Introduction and Use of Gunpowder

and Firearms among the Chinese," <u>Journal, Royal Asiatic Society, North-China Branch</u>, (1869-70), 73-104.
He thought that gunpowder may have been <u>imported</u> into China in the sixth century A.D.

176. NEEDHAM, Joseph. <u>Science and Civilization in China</u>. Cambridge, 1954 ff. (in progress).
This work, which includes technology, comes close to replacing all earlier publications on the topic. Vol. 5, which is actually to be five bound volumes, of which parts 2-4 have appeared, deals with "chemistry and chemical technology." Most of chemical technology, however, is to be in the unpublished parts 1 and 5, which are announced to deal with "martial technology" (including saltpeter), salt, textiles and ceramics. Glass has been covered in vol. 4, part 1. Profuse citations to both oriental and western sources.

177. PATEL, Bhulabhai. <u>Mineralien und Chemikalien der Indischen Pharmazie.</u> Braunschweig, 1963. (83 pp., a publication of the Pharmaziegeschictlichen Seminar, Technische Hochschule, Braunschweig)
Drugs "in der Ayurvedic und Unani-Therapie" are listed, analyzed, and compared with European equivalents. Citations.

178. RAHMAN, A. <u>Bibliography of Source Material on History of Science and Technology in Medieval India.</u> New Dehli, 1975.
An introduction to this publication was published in 1975 by the National Commission for the Compilation of History of Sciences in India, of the Indian National Science Academy. I have been unable to locate the publication itself.

179. REMUSAT, Abel, "Lettre sur l'existence de deux volcans brulans dans la Tartarie-Centrale." <u>AC</u>, ser. 2, 14 (1820), 309-11.
An early account of volcanic sal ammoniac, from a Chinese literary source. It is followed by comments by Louis Cordier.

180. SCHAFER, Edward. <u>The Golden Peaches of Samarkand</u>. Berkeley, 1963. (399 pp.)

An historical dictionary of imports to China from the west (i.e. central Asia and beyond). Citations.

181. SCHAFER, Edward. The Vermillion bird. Berkeley, 1967. (380 pp.)
An historical dictionary of imports into China from southeast Asia. Citations.

182. SMITH, E. Porter. "Chinese chemical Manufactures," Journal, Royal Asiatic Society, North China Branch, n.s. 6 (1870), 139-47.
A miscellany, designed "to show that there is such a thing as chemistry in China."

183. STUHLMANN, C.C. "Chinese soda," Journal, Peking Oriental Society, 3 (1895) 566
Not seen.

184. WANG LING, "On the Invention and Use of Gunpowder and Firearms in China," Isis, 37 (1947), 160-78.
Dates gunpowder from the early tenth century. Says that the Chinese emphasized its incendiary properties, the Europeans its propellant properties. Citations.

185. WHEATLY, Paul. "Geographical Notes on Some Commodities Involved in the Sung Maritime Trade," Journal, Royal Asiatic Society, Malayan Branch, 32 (1961), 2 ff.
Includes references to the importation of borax into China.

186. YAMAZAKI, T. "The Characteristic Development of Chemical Technology in Modern Japan, Chiefly in the Years Between the Two World wars," Tokyo University Studies in the History of Science and Technology, 1965.
Not seen.

187. YANG, Tzu-Chiu. "Chemical industry in Kuangtung province," Journal, Royal Asiatic Socity, North China Branch, n.s. 50 (1919), 133-43.
Describes traditional industries, rice wine, peanut oil, tung oil, pottery and porcelain, vermillion, red lead, soap.

188. ZWEHTKOFF, P. "Remarks on the production of salt in China," Journal, Royal Asiatic Society, North China Branch, n.s. 22 (1887), 81-87.

Remarkable as an attempt to give the historical evolution of salt-making methods in China. Translated, without citations, from a Russian version which I have not seen but suspect to have been more detailed.

AUSTRIA, HUNGARY, AND CZECHOSLOVAKIA

EARLY WORKS

189. SCHREYER, Josef. Kommerz, Fabriken und Manufakturen des K. Böhmen. Prague & Leipzig, 1790.
Various information on production of alum, vitriol, sulphur, potash, blue pigment (Blaufarben, smalt) and glass.

190. TOWSON, Robert. Travels in Hungary. London, 1797.
Describes production of salt, soda and saltpeter.

191. ANDRE, C.C. Neueste geographisch Statistische Bescreibung des Kaisertums Oesterreich. Weimar, 1813 (Bd. 15 of Neueste Lander Und Volkerkunde)
Describes the production of mercury, alum, saltpeter, potash, petroleum, smalt, white lead. Well indexed, with maps; this is one of the most informative books of its kind.

192. BRIGHT, Richard. Travels from Vienna through Lower Hungary. Edinburgh, 1818. (642 pp.)
Describes production of salt, sulphur, saltpeter, alum, and visits to the mineral productuion centers at Chemnitz and Kremnitz.

193. ELDELEN von KEES, Stephan. Darstellung des Fabriks und Gewerbwesens in österreichen Kaiserstaate. Vienna, 1823.
Pp. 944-1009 deal with "die chemischen Fabricate und Farben."

HISTORIES

194. BAUER, Alexander. "Chemische industrie," in W.F. Exner, ed., Beiträge zur Geschichte der Gewerbe und Erfindungen Oesterreichs. Vienna, 1873, 1:93-139.
Much information on factories and production data. Says that the Austrian chemical industry originated in

pyrites, their distillation for sulphur, weathering for vitriol, and the use of the latter to make oil of vitriol.

195. BUDAPEST, TECHNICAL UNIVERSITY. Hundred Years of the Faculty of Chemical Engineering, Technical University Budapest, 1871-1971. Budapest, 1972. (241 pp., in English)
Numerous authors, less concerned with history than the present situation.

196. DONATH, Edward. "Oesterreichs Antheil and der Entwicklung der chemischen Industrie," Oesterreichische Chemiker-Zeitung, 1905, Heft I-IV. (50 pp.)
Miscellaneous information on early factories.

197. GLASER, Julius. Die chemische Industrie Oesterreichs und ihre Entwicklung. Berlin, 1918. (77 pp.)
Largely economic history with bibliography but no citations. He says that the evolution of pharmacy into a chemical industry was rare.

198. HASSINGER, Herbert. "Der Stand der Manufakturen in den deutschen Erbländern der Hapsburgermonarchie am Ende des 18. Jahrhundert," Forschungen zur Sozial- und Wirtschaftsgeschichte, 6 (1964), 110-76.

199. HERBERT-KERCHNAWE, Ernst. Die Bleiweiss-Fabrication in Oesterreich. Vienna, 1898. (26 pp., Museum für Geschichte der österreichischen Arbeit, Monograph VIII)
The "Dutch process" was introduced at Klagenfurt in 1756.

200. MARTELL, Paul. "Zur Geschichte der chemische Industrie in Oesterreich," Chemische Industrie, 3 (1911), 205-10, 306-7.
Summary, without references.

201. MECHT, Otto. Die KK Spiegelfabrik zu Neuhaus in Neider-Oesterreich," Ein Beytrag zur Geschichte des Merkantilismus. Vienna, 1909. (164 pp.)
A dissertation notable for its careful documentation.

202. OTRUBA, Gustav. Die Wirtschaftspolitik Maria Theresias. Vienna, 1963.
Mentions various chemical works; says that every large city had a saltpeter works "to use the local urine."

203. SCHICHT, Heinrich. "The development and present situation of the chemical industry in Czechoslovakia," Journal, Society of Chemical Industry, 50 (1931), 276-77.
Actually deals with the prewar history, when the area was part of the Austro-Hungarian empire. Informative and not as brief as the page numbers suggest. It actually covers four pages. No citations.

204. SLOKAR, Johann. Geschichte der österreichischen Industrie. Vienna, 1914.
Chemicals dealt with on pp. 459-69. Citations.

205. TEICHOWA, Alice. "Industrielle Kartelle und die chemische Industrie der vormünchener Tschecko-slowaki," Tradition, 17 (1972), 143-72.
Assesses the influence, which the author finds to be considerable, of international cartels on the chemical industry of Czechoslovakia in the post World War I period. Citations.

206. WOAT, Thomas. Umriss der Entstehungs- und Entwicklungsgeschichte des fürstliche Auersperg'schen Mineralwerkes zu Gross-Lukavic in Böhmen. Gross-Lukavic, 1875. (16 pp.)
A capsule history of a chemical factory which began in the early 18th century with the exploitation of pyrites, made oil of vitriol from vitriol before 1750, English (i.e., chamber) sulphuric acid in 1807, and still existed at the date of publication.

207. WRANY, Adalbert. Geschichte der Chemie und der auf chemischer Grundlage beruhrenden Betriebe in Böhmen. Prague, 1902. (398 pp.)
An excellent, detailed history, with substantial attention to technology. Citations.

FRANCE

EARLY WORKS

208. CHAPTAL, J.A. Essai sur la perfectionment des arts chimiques en France. Paris, an. 8 (1799/1800). (88 pp.)
Why is France only second among the manufacturing nations of Europe? Because of a prejudice against the factory as a "métier object," of an administration which sees it only as a source of revenue, of a lack of "national spirit," and of "a scandalous taste for all foreign productions." He makes suggestions for rectifying all this.

209. SCHERER, J.A. "Ueber den jetzigen Zustand der Chemie in Frankreich und über die jetzigen Beschäftigung der dortigen Chemiker," Scherere's Allgemeine Journal der Chemie, 5 (1800), 123-24.
An optimistic report. The townsman (Burger), heretofore unable to occupy himself with factories, has brought his talents to bear, and chemical works are "more numerous than ever," especially for mineral acids, alum, vitriol, potash and sal ammoniac.

210. VIENNET, Odette, ed. Une enquête économique dans la France impériale: le voyage du hambourgeois Philippe-Andre Nemnich, 1809. Paris, 1947.
Trans. from Nemnich's Beiträge zur eigentliche kenntnis von Frankreich, which was vol. 5 of his Tagebuch einer der Kultur und Industrie gewidmeten Reise (Tübingan, 1809-11. 8 vols.). Much information on all aspects of the French chemical industry and trade.

211. CHAPTAL, J.A. De l'industrie francaise. Paris, 1819. (2 vols)
Chaptal was not only a notable chemist and chemical manufacturer but was at this time Minister of the Interior of France. He is now more optimistic than he had been in 1800 (no. 208), thinks that France has regained its lead. Much on chemical industry.

212. (chemicals exhibited at the Louvre in 1823). Bulletin des Sciences Technologiques, 1 (1824), 73-76.

Lists chemical exhibitors and the materials displayed (artificial soda, borax, sal ammoniac, Glauber's salt, etc.) at an early "trade show."

213. "Coup d'oeil sur l'état actuel de l'industrie manufacturière en France. Arts chimiques." <u>BSE</u>, 24 (1825), 84-90, 152-62, 186-95.
Lists firms and products.

HISTORIES

214. BAUD, Paul. "Les débuts de l'industrie chimique en France," <u>Annales de l'Université de Paris</u>, 7 (1932), 223-41.
Not seen.

215. BAUD, Paul. <u>L'Industrie chimique en France: études historique et géographique</u>. Paris, 1932. (418 pp.)
Part I deals with traditional industries; Part 2 with the "grande industrie," and includes a chapter on each region in France. Citations.

216. BAUD, Paul. "Les origines de la grande industrie chimique en France," <u>Revue Historique</u>, 174 (1934), 1-18.
A description of some "familial" chemical industries before the Revolution. Citations. An important article.

217. BEDHOME, R. "L'Evolution de la carbochimie dans le Basin du Nord et du Pas-de-Calais," <u>Chimie Industrielle</u>, 87 (1962), 435-39.
Dates the effective industry from the recovery of coke by-products, especially ammonia, in 1905; a second era after 1945, when the mines were nationalized. An undocumented but informative reminiscence by a participant.

218. CHAGNON, A., P. COSTE, and M. LACOIN. <u>Les débuts de la grande industrie chimique et la Société d'Encouragement pour l'Industrie Nationale</u>. Paris, 1945. (38 pp.)
Three articles, "Historique du procédé Leblanc," by Chagnon, "Le procédé Leblanc et l'industrie chimique au siècle dernier," by Coste, and "Chaptal, Minèstre

de la production industrielle du Premier Consul," by Lacoin.

219. DAVY, R. Contribution à l'étude des origines de la droguerie pharmaceutique et de l'industrie du sel ammoniac en France: L'apothecaire Antoine Baumé (1728-1804). Cahors: diss., Univ. of Strassbourg, 1955.
Not seen.

220. FAUQUE, Maurice. L'évolution économique de la grande industrie chimique en France. diss., Univ. of Strassbourg, 1932. (251 pp.)
Not seen.

221. GILLISPIE, Charles C. "The natural history of industry," Isis, 48 (1957), 398-407.
Using France as an example, addresses the question of when "scientifically instructed entrepreneurs" took control of the chemical industry from "gothic mastercraftsmen of olden times." Dates it from the Revolution. Citations.

222. KERSAINT, G. "Sur la fabrique de produits chimiques établie par Fourcroy et Vauquelin, 23, rue de Colombier à Paris," CR, 247 (1958), 461-4, and Revue d'Histoire de la Pharmacie, 47 (1959), 25-30.
Describes the chemical factory established in an. xiv (1805/06) adjoining the church of St. Germain de Pres. Drawing of the site.

223. KOLBE, J. Sur l'évolution actuelle de la grande industrie chimique. Lille, 1883.
Not seen.

224. SMITH, John Graham. The Origins and Early Development of the Heavy Chemical Industry in France. Oxford: Clarendon, 1979. (396 pp.)
Although limited in scope--to sulphuric acid, chlorine bleaching, Leblanc soda, and to the approximate period 1760-1820--this book is an outstanding exemplar of the possibilities of serious historical research in industrial chemistry. Citations, many from archival sources.

225. THEPOT, André. "Le système continental et les débuts de l'industrie chimique en France," Revue de l'Institut Napoléon, 99 (1966), 79-84.
Except for the rupture with Spain, which led the government to favor the artificial soda industry, the Continental Block had little immediate effect on the large-scale chemical industry which was coming into existence at the beginning of the "Empire." Citations.

226. TURGAN, J.F. Les grandes usines de France. Tableau de l'industrie français au xixe siècle. Paris, 1860-1870. (9 vols)
An elegantly illustrated popular description of French and some foreign industries with an historical bias. Chemical works described include Marseilles' soap, St Gobain glass, Murano glass, beet sugar, alcoholic beverages, aluminum, rubber, aniline dyes, and the Paris gasworks.

GERMANY

EARLY WORKS

227. GLAUBER, J.R. Prosperitatis germaniae. Amsterdam, 1656.
A prescription for national salvation through chemistry, from one who deserves, as much as any, to be called the "father" of industrial chemistry. It is analyzed by Alfons Kotowski, in "Deutschlands Wohlfahrt," Glauber's Gedanken über die Hebung des deutschen Nationalreichtums durch die Chemie," ZAC, 52 (1939), 109-12.

228. ERXLEBEN, J.C.P. "Nachrichten von einer Reise nach dem Weisner," Hannovarisches Magazin, 3 (1765), 994-1023.
Report on a field trip, by a Göttingen student (subsequently a distinguished scientist) to an early "chemical complex" near Allendorf (Hesse). Describes the state saltworks, a privately owned alum works, and a factory making mysterious clay vessels destined for shipment to Holland.

229. "Nachricht von den Gravenhorstischen chymischen Producten," Berliner Sammlung zur Beförderung der

Arzneiwissenschaften, 5 (1773), 148-55.
Says that the Gravenhorst factory at Braunschweig produced sal ammoniac, red alum, green pigment (Braunschweigisches Grün), Glauber's salt, artificial alkali (from salt), and a medical soap. The author wishes that the Gravenhorsts, who were notorious for secrecy, would tell more (to silence those who criticize them). But they were also notorious for publishing small articles praising their products (without revealing much about them), and this reads like a summary of those publications, which were little more than advertisements. (See also no. 708.)

230. "Reise von Berlin nach Strasburg," in (Johann Bernouilli's) Sammlung kurzer Reisebeschreibungen, 3 (1781), 135-70.
This unidentified traveler was interested in chemistry and visited vitriol, sal ammoniac, and saltpeter works in Magdeburg and Goslar (Rammelsberg).

231. GATTERER, C.W.J. Anleitung den Harz und andere Bergwerke. Göttingen, 1785. (2 vols)
This important book includes descriptions of chemical products obtained from mercury, antimony, bismuth, zinc, cobalt, arsenic, lead, gypsum, lime, and "salts," each with a bibliography.

232. HEINTIZ, F.A. von. Abhandlung über die Produkte des Mineralreiches in den Königliche-Prussischen Staaten. Berlin, 1786. (113 pp.)
A detailed description of the mineral industries of Prussia at that time. He says that the king began to take a direct interest in this resource in 1753. There was also a French edition.

233. "Bemerkungen...einer Reise nach Niedersachsen," Bergmannisches Journal, 2 (1794), 260-78, 282-307.
Recounts visits to the Goslar vitriol works and the saltpeter refinery at Rothenburg.

234. "Übersicht von dem Fabrik und Manufakturwesens in den Frankischen Fürstenthümern Bayreuth und Anspach," Handlungszeitung, 15 (1798), 1-2, 9-11, 17-19, 25-28.
An unusually complete and detailed account, with statistics, of the manufactures of a region, including

alum, vitriol, potash, saltpeter, and the pigment Prussian blue.

235. DUHAMEL, J.P.F. "Aperçu des richesses minerales... Dept. de la Sarre," <u>Journal des Mines</u>, 15 (1804), 321-26.
Reports on the manufacture of alum, sal ammoniac, magnesium, sulphate and Prussian blue in the Saar region made while it was under French occupation.

236. ADAMS, John Quincy. <u>Letters on Silesia</u>. London, 1804.
The subsequent President of the United States reports, among other things, production methods for vitriol, sulphuric acid, and quicklime in Silesia.

237. SUCKOW, G.A. <u>Bemerkungen über einige chymische Gewerbe</u>. Mannheim, 1809. (36 pp.)
Advocates domestic production of "imported" commodities, namely tartar, verdegris, white lead, sugar of lead, brandy, oils, chlorine bleach, alum and Epsom salt. Since most, if not all, of these were then produced in Germany, he probably refers to his own state, Baden.

HISTORIES

238. BAAR, Lother. <u>Die Berliner Industrie in der industriellen Revolution</u>. Berlin, 1966.
Mentions the chemical factories of Riedel (1814) and Kunheim (1826) but finds that the chemical industry was of little importance before 1860, and the heavy chemical industry never of much importance because of the need of importing raw materials.

239. BENAERTS, P. <u>Les origines de la grande industrie allemande</u>. Paris, 1933.
Deals mostly with the period of the Zollverein, with background material on the earlier period 1815-34. The chemical industry is important.

240. BINZ, Arthur, "Entwicklung der chemische Technologie in Deutschland," <u>ZAC</u>, 49 (1936), 707-09.
Not seen.

241. BORSCHEID, Peter. <u>Naturwissenschaft, Staat, und</u>

Industrie in Baden, 1848-1914. Stuttgart: Klett, 1976. (242 pp.)
The chemical industry is prominent. Citations.

242. CARO, Heinrich. "Ueber die Entwicklung der chemischen Industrie von Manheim-Ludwigshafen," ZAC, 17 (1904), 1343-62.
An excellent documented summary history of one of the most important German chemical-producing regions.

243. CHRISTIANSEN, Carl C. Geschichte des Standorts der chemische Industrie. Tübingen, 1913. (44 pp.)
A dissertation, on circumstances controlling the location of chemical industries, especially in Germany. Citations.

244. FECHNER, H. Wirtschaftsgeschichte der Preussischen Provinz Schlesien zur Zeit ihrer provinziellen Selbständigkeit, 1741-1806. Breslau, 1907.
Provides miscellaneous information on the production of saltpeter, nitric and sulphuric acids, Prussian blue, and verdegris (Grünspan).

245. FORBERGER, Rudolph. Die Manufaktur in Sachsen von Ende des 16 bis zum Anfang des 19 Jahrhunderts. Berlin, 1958. (456 pp.)
Miscellaneous information on the production of gunpowder, cobalt pigments (Blaufarben), white lead, verdegris.

246. HABER, Fritz. "Die deutsche Chemie in der letze 10 Jahren," in his Aus Leben und Beruf, Berlin, 1927, 7-24.
An address to the German Club of Buenos Aires, lamenting the postwar state of Germany, blaming it on the victors of 1918. Credits the growth of the German chemical industry to research.

247. HOECHST (firm). US-Administration: die Verwaltung des Werkes Hoechst, 1945-1953. Frankfurt/M, 1976. (153 pp. DHA no. 48)
Documents, with a brief introduction, beginning with an order from the American Military Govt. of Germany, to the civil population (Apr. 4, 1945), and ending with another order releasing Hoechst from allied control (March 27, 1953).

248. JACOB, Stefan. Chemische Vor- und Frühindustrie in Franken. Düsseldorf, 1968. (411 pp.)
A dissertation with extensive documentation. The period, seventeenth to early nineteen century; the materials, saltpeter, potash, alum, vitriol, cobalt colors, white lead, Prussian blue; the geographic area, Franconia, centered on Frankfurt/M and was enclosed by Lower Saxony, Thuringia, Bavaria and the French and Belgian borders.

249. JOHNSON, Jeffery A. "The Chemical Reichsanstalt Association; Big Science in Imperial Germany." Dissertation, Princeton University, 1980.
Not seen.

250. KUNZMANN, T. Die Bedeutung der wissenschaftlichen Tätigkeit von J. von Liebig, F. Wöhler, und C.F. Schönbein für die Entwicklung der deutschen chemischen Industrie. Berlin, 1931 (90 pp.)
Not seen.

251. LAAR, J. et al. "Beiträge zur Geschichte der pharmazeutischen Industrie in Deutschland," Forschung. Praxis. Fortbildung, 17 (1966) 558-600.
A brief introduction is followed by capsule histories of 24 German firms. No references.

252. LEPSIUS, B. Deutschlands chemische Industrie, 1888-1913. Berlin, 1914. (64 pp.)
A concise but informative survey written as part of a celebration of the 25th anniversary of the accession of Kaiser Wilhelm II.

253. MARTELL, Paul. "Zur Geschichte der chemischen Industrie in Schlesien," Chemischen Industrie, 34 (1911), 4-9.
Brief and without references but packed with information.

254. MARTELL, P. "Zur Geschichte der chemische Industrie in der Mark Brandenburg," Chemische Industrie, 35 (1912), 2 ff
Not seen.

255. RUSKE, Walter. "Wirtschaftspolitik, Unternehmertum

und Wissenschaft am Beispiel der chemischen Industrie
Berlins im 19. Jahrhundret," <u>Technikgeschichte in
Einzeldarstellungen</u>, 16 (1970), 81-112.
A meticulous description and analysis of the chemical
industry in and around Berlin, emphasizing personali-
ties and institutions which he sees as representing
the ascent of liberal economics and decline of
mercantilism.

256. SCHALL, Horst. <u>Der chemische Industrie Deutschlands</u>.
Nurnburg: Hochschule für Wirtschafte und Solzial-
wissenchaftlichen Nurnbergs, 1959. (153 pp.)
On factors in the location and organization of the
chemical industry.

257. SCHOENEMANN, Karl. "The Separate Development of Chemical Engineering in Germany," in <u>Furter</u> (1980, no.
116), 249-72.
Says that the German chemical industry "exclusively
employed chemists" prior to the ammonia synthesis and
that "chemical engineering in the proper sense did not
appear any sooner than the middle 1960's." Citations.

258. SCHRAPS, - <u>Der Drogenhandel en gros und en detail in
Deutschland</u>. Leipzig, 1904.
Not seen.

259. SCHULTZ, Gustav. "Die chemische Industrie Bayerns,"
in <u>Darstellung aus der Geschichte der Technik...in
Bayern</u>, Munich, 1906.
Not seen. This is cited by Fester (no. 111)

260. SCHULZE, Hermann. <u>Die Entwicklung der chemische Industrie in Deutschland</u>. Halle, 1907. (159 pp.)
"Ein volkswirtschaftliche Studie." A dissertation,
discussing various branches of the industry and its
cartellization.

261. SCHULZE, Hermann. <u>Die Entwicklung der chemischen
Industrie in Deutschland seit dem Jahre 1875</u>.
Halle, 1908. (309 pp.)
Various data, from 1849, on factories and production.

262. SCHWERIN VON KROSIGK, L. von. <u>Die grosse Zeit des</u>

Feuers. Tübingen, 1957. (3 vols)
An immense (2,130 pp.) history of German industry, organized around biographies of scientists, engineers, industrialists, labor leaders and politicians. A personal account without citations but indexed and with a good bibliography. Much on the chemical industry. Presumably the product of the forced leisure of the author, who was German Finance Minister from 1932 to 1945.

263. SPETER, Max. "Die 'chymischen Fabriken von Deutschland' um 1799." Chemiker Zeitung, 56 (1932), 391-92. As reported in J.C. Gädicke's Fabriken und Manufacfacturen Address-Lexicon (1799). Useful notes.

264. STEINBECK, Aemil. Geschichte des schlesische Bergbaues. Breslau, 1857. (2 vols)
Contains histories of the production of saltpeter, alum and vitriol.

265. TROITZCSH, Ulrich. Ansätze technologischen Denkens bei den Kameralisten des 17. und 18. Jahrhunderts. Berlin, 1966. (193 pp.)
The intellectual privy-councillors of Germany--the Cameralists--were prolific authors on many subjects, including chemical technology. This is a guide to their works, including some authors (e.g., Beckmann, Justi, Zincke) cited in this bibliography.

266. VOGEL, Friedrich M. Die Entwicklung des Exports und Imports der chemische Industrie in Deutschland. Munich, 1914. (146 pp.)
Much data, with a general indication of sources. A dissertation at the University of Erlangen.

267. WALTER, Gustav A. Die Geschlichtliche Entwicklung der rheinischen Mineralfarben-Industrie. Essen, 1922. (204 pp.)
From the beginning of the 19th century until 1914. Begins with a general discussion of the pigments (white lead, zinc white, smalt, ultramarine, Prussian blue, etc.) and follows with histories of the companies involved. Citations and bibliography. A model history.

268. WELSCH, Fritz. "Zur Entstehung der Sodaindustrie in Deutschland," NTM, 9 (1972), 49-52.
Cites the production of artificial soda at Schönebeck saltworks about 1802, and claims this to be the first Leblanc works outside of France. Citations.

269. WITT, O.N. Die chemische Industrie des deutschen Reiches im Beginne des 20. Jahrhunderts. Berlin, 1902. (228 pp.)
Not seen.

270. WITT, O.N. "Die Entwicklung der deutschen chemische Industrie im 19. Jahrhundret," Chemische Industrie, 26 (1903), 149-58.
An address. Claims that the German chemical industry rests not on "Empire" but on "exact research." Notable for its freedom from nationalistic hyperbole.

GREAT BRITAIN

EARLY WORKS

271. TAUBE, F.W. von. Abschilderung der Engländischen Manufacturen, Handlung, Schiffart und Colonien. Th. 1, Vienna, 1777. (231 pp., 2nd ed., the first dated 1774)
Includes a detailed discussion of chemical commodities, their qualities, prices, foreign competition, the use of coal as fuel, and how the English survived the wartime loss of American potash.

272. FABRICIUS, J.C. "Mineralogische und Technologische Bemerkungen auf einer Reise durch verschiedene Provinzen in England und Schottland," in J.J. Ferber (1778, no. 109). 401-62.
Miscellaneous information, organized by counties, but little on chemical industries, except salt.

273. FAUJAS SAINT-FOND, B. Voyages en Angleterre, en Ecosse et aux isle Hébrides. Paris, 1797. (2 vols. An English translation was published in London, 1799)
This French traveler made many perspicacious observations as on the English use of coal: In France the poor simply stay in bed in winter while in England they work, crowded around their coal fires.

Prohibited from visiting the great sulphuric acid works at Prestonpans, which was surrounded by a wall, he analyzed its operation through "the suffocating smell perceived at a distance."

274. SVEDENSTJERNA, E.T. Reise durch einen Theil von England und Schottland, in den Jahren 1802 und 1803. Marburg, 1811. (194 pp., English trans., Newton Abbot: David and Charles, 1973)
Little on chemistry, but what there is is important. He visited Dundonald's coal tar distillery and Mackintosh's alum and vitriol works, among others.

275. DODD, George. Chemicals. London, 1844. (248 pp.)
This is vol. I in a series on British Manufacturers (1844-51). Describes the state of the art in Britain, dealing with sulphur, salt, soap, potash, candles, oils and resins, porcelain and pottery, colors, starch, "street gas." Interesting illustrations.

HISTORIES

276. The Fight for Supremacy. Liverpool, 1907, (77 pp.)
Reprints articles from the Times, Nov. 2-26, 1906, giving a brief history of the British alkali industry up to the formation of the United Alkali Co., and the dimensions of its problem in maintaining "supremacy." No references.

277. ARMSTRONG, W.,G. et al. The Industrial Resources of the Tyne, Wear and Tees. London, 1864. (361 pp.)
About 80 pages are devoted to ceramics, paper, tanning, and the chemical industry, including a detailed account of late eighteenth century attempts to make artificial soda, by William Losh, Thos. Doubleday, and others at Newcastle. No citations.

278. BARKER, T.C. and J.R. Harris. A Merseyside Town in the Industrial Revolution. St. Helens, 1750-1900. Liverpool, 1945.
A detailed and fascinating history of a British chemical town. Citations.

279. CAMPBELL, W.A. The Old Tyneside Chemical Trade. Newcastle upon Tyne, n.d. (ca 1961). (62 pp.)

A brief, undocumented, but very informative and well-illustrated account of one of the first and most important centers of the British chemical industry.

280. CLOW, Archibald. "The Timber Famine and the Development of Technology," AS, 12 (1965), 85-102.
Deals primarily with the alkali industry in Britain. Citations.

281. FULMER, June. "Technology, chemistry and the law in early 19th century England," T&C 21 (1980), 1-28.
Extracts, from an 1819 lawsuit between a sugar refiner and his insurer, evidence for the conditions and impact of of the chemical industry at the time. Citations.

282. GIBBS, F.W. "Peter Shaw and the Revival of Chemistry," AS, 7 (1951), 211-37.
An important account of mid-18th century attempts to promote industrial chemistry in England. Citations.

283. HALL, Marie Boas, "La croissance de l'Industrie chimique en Grande-Bretagne au xix siècle," Revue d'Histoire des Science, 26 (1973), 49-68.
Says that the industry depended not upon science but on "industrial and mechanical imagination." Dates its apogee about 1870. Citations.

284. HARDIE, D.W.F. A History of the Chemical Industry in Widnes. Birmingham, 1950. (250 pp.)
From the foundation of an alkali works in 1847 to the present. A remarkably detached and instructional history of one of the most notorious "alkali towns" (a map shows 28 chemical works existing during "the Leblanc period"). Few citations, but useful appendices. Largely based on "oral tradition."

285. HARDIE, D.W.F. "The Muspratts and the British chemical industry," Endeavor, 14 (1955), 29-33.
Deals with the careers of James Muspratt, "the founder of the British heavy chemical industry," and of James Sheridan Muspratt. Citations.

286. HARDIE, D.W.F. "The chemical industry in Merseyside," Proc., Chemical Society (London), (1961), Feb. 52-57.

Undocumented but informative summary history of the principal region of the British chemical industry.

287. HARDIE, D.W.F. and J.D. PRATT. A History of the Modern British Chemical Industry. London, 1966. (380 pp.)
Begins with "the chemical phase of the industrial revolution (1750-1850)." Especially valuable for new (twentieth century) industries. Includes brief company histories. No references.

288. HESSE, Albert, and H. GROSSMANN, eds. England's Handelskrieg und die chemische Industrie. Stuttgart, 1917. (344 pp.)
Lists itself as Band 2 of a work of the same title published in 1915 (which I've not seen) although it may be a revision. It is rather a source book than a history, as it consists of German translations of English, French and American articles exhorting their countrymen to overcome the German lead in chemistry.

289. HUME, John R. "The Saint Rollox Chemical Works," Journal of Industrial Archeology, 3 (1966), 185-92.
A historical sketch of this famous chemical works, established near Glasgow in 1799. The works was demolished in 1964-65, the oldest surviving building being an office of the nineteenth century.

290. LISCHKA, Johannes R. "Ludwig Mond and the British Alkali Industry. Dissertation, Duke University, 1975. (235 pp. Univ. Microfilm 71-10, 392)
Primarily concerned with the period 1870-90, when Mond introduced the Solvay process and contested with the established Leblanc industry. Citations, including manuscript materials from the archives of ICI.

291. MACTEAR, James. "On the growth of the alkali and bleaching-powder manufacture of the Glasgow district," Chemical News, 35 (1877) 4-5, 14-17.
This "slight historical sketch" is nevertheless informative, especially on economic and production data.

292. MIALL, Stephen. History of the British Chemical Industry. London, 1931. (273 pp.)
A summary account, written for the 50th anniversary of the Society of Chemical Industry. Useful for capsule

histories of firms and for chronological tables.

293. MORGAN, G.T., and D.D. PRATT. British Chemical Industry: Its Rise and Development. London, 1938. (400 pp.)
Despite the title this is largely devoted to technical descriptions of the state of the art. No references.

294. MUSSEN, A.E., and Eric ROBINSON. Science and Technology in the Industrial Revolution. Manchester, 1969. (534 pp.)
Restricted to Britain. The chemical industry is treated pp. 231-371, notably alkali, chlorine bleaching, dyestuffs, and the activities of the Henry family of Manchester. Citations.

295. NEF, John U. Rise of the British Coal Industry. London, 1932. (2 vols)
The fundamental work on this topic, it gives substantial attention to the chemical industries, which accounted for much of the demand for coal. Citations.

296. NEF, John U. "The Progress of Technology and the Growth of Large-Scale Industry in Great Britain, 1540-1640," Economic History Review, 8 (1934), 3-24.
Sees an earlier industrial revolution, about 1540-1660, characterized by the progress of technology and the development of a coal-burning economy. Emphasizes new industries of the late sixteenth century, paper, gunpowder, cannon, alum and copperas (vitriol), sugar, saltpeter, brass. Citations.

297. PADLEY, Richard. "The Beginnings of the British Alkali Industry," Univ. of Birmingham Historical Journal, 3 (1951), 64-78.
A detailed account of attempts to manufacture artificial alkali in Britain prior to James Muspratt's Liverpool factory of 1823. Citations.

298. PARKER, Alice, "Samuel Skey (1726-1800) and Chemical Industry in Bewdley," TNS, 27 (1951), 217-18.
A dye-maker, Skey learned the lead-chamber process (for sulphuric acid) from a workman who absconded from Roebuck's Prestonpans works. Skey became the second British manufacturer of the acid by this process. Much detail, despite its brevity.

299. RICHARDSON, H.W. "The Development of the British Dyestuffs Industry before 1939," <u>Scottish Journal of Political Economy</u>, 9 (1962), 110-29.
Economic history, exploring Britain's failure to develop a substantial dye industry before 1914, its progress with state support thereafter--"the first conspicuous example of an industry in Britain receiving extensive state tutelage." Concludes that it was justified but "whether or not continued protection is justified since 1945 is another question altogether." Citations.

300. RICHARDSON, H.W. "Chemicals," in D.H. Aldcroft ed., <u>The Development of British Industry and Foreign Competition, 1875-1914</u>. Toronto, 1968: 307-25.
Concludes that, qualifications nothwithstanding, "the British chemical industry before 1914 was the first, but by no means the last, victim of the sin of underestimating the returns from research and development. But the sinner was not so much the British chemical master as society at large." Citations.

301. SALZMANN, L.F. <u>English Industries of the Middle Ages</u>. Oxford, 1923. (359 pp.)
Includes coal, pottery, glass, leather and brewing. Citations.

302. SCHINCK, E.R., Angus SMITH, et al. "Report on the Progress of Manufacturing Chemistry in South Lancashire," <u>BAAS Annual Rept.</u> (1861), 108-28.
Covers only "the past ten years" but with information of significant historical interest.

303. SCHOFIELD, M. "Roebuck and Keir," <u>Chemistry and Industry</u>, (1946), 106-07.
Some particulars on the enterprises of two early chemical manufacturers, John Roebuck (1718-94) and James Kier (1735-1820).

304. TENNANT, E.W.D. "The Early History of the Saint Rollox Chemical Works," <u>Chemistry and Industry</u>, (1947), 667-73.
A circumstantial account with quotations from documents, by the fifth generation of the proprietors, the Tennant family.

305. THURSTON, E.F. Winnington Research Laboratory, the first 50 years. n.p. (Winnington ?), n.d. (1978 ?) (56 pp.)
History of the laboratory opened by Bruner, Mond & Co. in 1928, after it had become (1927) part of the newly formed ICI (by which this was probably published).

306. WARREN, Kenneth. Chemical Foundations: the Alkali Industry in Britain to 1926. Oxford: Clarendon, 1980. (208 pp.)
Particularly concerned with industrial geography. A detailed history with numerous citations in the text, many to the archives of ICI.

307. WHITTAKER, C.M. "Some Early Stages in the Renaissance of the British Dyemaking Industry; Tales from Turnbridge. Huddersfield, 1899-1920." Journal, Society of Dyers and Coulorists, 72 (1956), 557-63.
Personal reminiscences of one employed in the industry from 1899, a time when "chemists in dye-houses were the exception and were not beloved by the foreman dyers." The British industry was then at "rock bottom." Very informative. A few references to patents.

308. WOODRUFF, Wm. The Rise of the British Rubber Industry during the 19th Century. Liverpool, 1958. (246 pp.)
A detailed economic history with much attention to technology. Citations.

HOLLAND and BELGIUM

The curious state of the history of chemical technology is here exemplified, for there appears to be no history of the Dutch chemical industry, the earliest (perhaps after that of Venice) of any importance in Europe; while there is an excellent history of the early chemical industry of Belgium, of which very little contemporaneous notice was ever taken.

EARLY WORKS

309. FERBER, J.J. "Nachrichten von einigen chymischen Fabriken," in his Neue Beytrage zur Mineralgeschichte, Mittau, 1778, Ed. 1: 315-90.
Describes the production of sulphur, aqua fortis, cinnabar, borax, corrosive sublimate, camphor and drugs at Amsterdam and Rotterdam.

310. VOLKMANN, J.J. Neueste Reise durch die Vereinigten Niederlände. Leipzig, 1783. (2 vols)
"Assembled from the best reports and newest writings" --this is not first hand. He says that Dutch trade (and manufacturers) have declined continuously since the Seven Years War (1756-63); but he includes considerable information on them.

311. EVERSMANN, F.A.A. Technologische Bermerkungen auf einer Reise durch Holland. Freyberg, 1792. (236 pp.)
Gives first-hand descriptions of factories for cobalt colors, cinnabar and other mercurical products, white lead, sea salt, quicklime and nitric acid. Much of the book was published without attribution in Bergmännisches Journal, 2 (1791) 72-109, 279-308, 329-47.

312. NEMNICH, P.A. Original-Beiträge zur eigenlichen kenntniss von Holland. Tübingen, 1809 (2 vols) (being vols. 3 and 4 of his Tagebuch einer der Kultur und Industrie gewidmeten Reise).
This contains the most complete description of the Dutch chemical industry known to me based on a visit made in 1809-09.

313. UILKINS, J.A. Technologisch Handboek. Amsterdam, 1809-19. (3 vols)
Mostly from foreign sources but includes remarks on the present state of the Dutch chemical industry.

314. "Ueber den Zustand der Industrie in nördliches Holland," (Dingler's) Polytechnisches Journal, 37 (1830), 225-27.
Says that the Dutch chemical industry is still important, despite its misfortunes and supports this through examples--the French are obliged to sell their white lead under the name "Dutch white lead," etc.

315. BAUMHAUER, E.H. van. Het nut der scheikunde voor den industrie. Haarlem, 1855 (39 pp.)
A lament, reporting the Dutch chemical industry to be behind that of Germany, Belgium, England and France and explaining it mainly by blaming the Dutch neglect of chemical science.

HISTORIES

316. ANDRE-FELIX, Annette. <u>Les débuts de l'industrie chimique dans les Pays Bas autrichiens</u>. Brussels: Universitaire Libre, Institut de Sociologie, 1971 (148 pp.)
Detailed account with particular reference to business firms, of the production of sal ammoniac, sulphuric and nitric acids in Belgium in the eighteenth century. Citations, mostly to unprinted sources. A model for similar histories of other regions.

317. WOLTERECK, Heinz. <u>Die Entwicklung der chemischen Industrie Hollands in den Jahren 1914 bis 1925</u>. Leipzig, 1927. (95 pp.)
A dissertation with citations.

ITALY

EARLY WORKS

318. MERCATI, Michele. <u>Metallotheca</u>. Rome, 1719. (378 pp.)
Deals with alum, sal ammoniac, sulphur, atrament (vitriol) and other materials in relation to Italian practice. A very posthumous publication edited by J.M. Lancisi. Mercati (d. 1585) was chief physician to Pope Clement VIII.

319. FERBER, J.J. <u>Travels in Italy</u>. London, 1776. (377 pp. trans. of his <u>Briefe aus Wälshland</u>, Prague, 1773)
Letters reporting a voyage in 1771, and visits to the alum works at Solfaterra (near Naples) and Tolfa (in the Papal States), as well as the mercury mines in Idria.

HISTORIES

320. ASTARITA, Gianni. "The History of Chemical Engineering in Italy," In <u>Furter</u> (1980, no. 116), 205-26.
Chronological, beginning with the Middle Ages. Discusses engineering schools at Naples, Rome, Milan, Bologna, Padua and Palermo; but the first university to give any prominence to chemical engineering seems

to have been Naples in 1901. Comes up to date. Citations.

321. GINORI-CONTI, P. "L'Industrie chimique in Italie," <u>Chemie et Industrie</u>, special issue (1929) 728 ff.
Not seen.

322. KOERNER, -. <u>L'Industrie chimico in Italie nel cinquantennio 1861-1910</u>. Milan, 1911.
Not seen.

323. LORIA, Mario. <u>Camillo Cavour e l'industrie chimica dei concimi</u>. Torino, 1964. (111 pp.)
On the fertilizer industry of Piedmont in the nineteenth century from guano to superphosphates. Brief text with numerous documentary appendices. See also his "Cavour and the development of the fertilizer industry in Piedmont," <u>T&C</u>, 8 (1967), 159-77.

324. RIENZI, Emanuele. <u>L'Industria dell'acido solforico in Italia</u>. Rome, 1940. (344 pp)
A chronological history of sulphuric acid, with special reference to Italy, occupies pp 11-48. References.

325. SQUARZINA, F. <u>Produzione e commercio dello zolfo in Sicilia nel secolo XIX</u>. Turin, 1963.
Not seen.

RUSSIA and POLAND

EARLY WORKS

326. PALLAS, P.S. <u>Reise durch verschiedene Provinzen des russischen Reichs</u>. St. Petersburg, 1771-76. (3 vols)
Includes descriptions of the production of salt, saltpeter, vitriol, alum and potash.

327. GEORGI, J.G. <u>Bemerkungen einer Reise im Russischen Reich im Jahre 1772</u>. St. Petersburg, 1772.
Describes factories for vitriol and white lead.

328. TOOKE, William. <u>View of the Russian Empire</u>. London, 1799. (3 vols)
Gives statistics on importation, exportation and

production of salt, potash, saltpeter, alum, vitriol, and mercury.

329. PETRI, J.C. "Ueber Russlands einheimische Natur-Produkte," *Journal für Fabrik*, 20 (1801), 177-213.
Few chemicals mentioned. "There are easier ways to make money in Russia than in factories, and this is the main reason why they are few and small."

330. PALLAS, P.S. *Travels through the Southern Provinces of the Russian Empire*, 1793-94. London, 1802-03. (2 vols)
Extends his observations (begun in no. 326) to other parts of the country.

HISTORIES

331. FUHRMANN, J.T. *The Origins of Capitalism in Russia*. Chicago: Quadrangle, 1972.
Includes references to chemical industries, which are said to have been mainly owned or operated by foreigners (Dutch and English).

332. MENDELEEV, D.I. "Chemical Industries," in Russia, Dept. of Trade and Manufacture, *Industries of Russia*. St. Petersburg, 1893, vols 1 & 2: 225-38. (prepared for the Columbian Exposition of 1892)
An informative but regrettably brief account by one of the great chemists of the age.

333. PANTYUCHOFF, N. "Ten Years of Russian Chemical Industry, 1898-1908," *Chemical Trade Journal*, 13 (1910), 63-65.
The "meagre details available" are sufficient for this substantial description of an industry which was new and nearly restricted to heavy chemicals. Statistics. No references.

334. SARNEKI, K. "Précis d'histoire de la technologie chimique inorganique en l'ancienne Pologne," *Actes, 11th International Congress of the History of Science*, Warsaw-Cracow, 1968, 4: 136-42.
Scattered references to a modest but not insignificant industry (the potash works established at Jamestown, Virginia, in 1609, were directed by a Pole). The first sulphuric acid works were at Warsaw in 1822.

335. SULER, B. "Ueber den gegenwärtigen Stand der anorganischen chemischen Industrie in Russland," in <u>Berichte, V. International Kongress für angewandte Chemie</u> (Berlin, 1903) 1: 746-55.
Useful, especially for statistics.

SPAIN, PORTUGAL, & LATIN AMERICA

EARLY WORKS

336. ULLOA, Antonio de. <u>Voyage to South America</u>. Dublin, 2nd ed., 1765. (2 vols., from Spanish ed. of 1748)
Reports observations made 1735-46 in the course of the French geodetic expedition on the production of sulphur, vitriol, guano, and mercury. No reference is made to the (Chili) nitrate deposits.

337. BOWLES, William. <u>Introduction à l'histoire naturelle et la geographie physique d'Espagne</u>. Paris, 1776. (516 pp.)
His observations include visits to sites of production of alum, salt, saltpeter, mercury, and cobalt.

338. TOWNSEND, Joseph. <u>A Journey through Spain</u>. London, 1792. (3 vols)
This book is remarkable for its extensive observations on technology, including the production of salt, saltpeter, gypsum, mercury, and soda.

HISTORIES

339. "La produccion quimica Espanola. Su evolucion indusdrial y sus perspectivas en Espana," in Banco de Bilbao. <u>Un siglo en la vida del Banco de Bilbao</u>. Bilbao, 1957, 332-92.
An historical summary, item by item, of the chemical industry of Spain, with few dates, but dealing mostly with the twentieth century. No references.

340. BUSTELO VAZQUEZ, Francisco. "Notas y commentarios sobre los origines de la industria espanola del nitrogeno," <u>Moneda y Crédito</u>, 63 (1957), 23-40.
Relates the checquered history of attempts to establish a synthetic nitrate industry in Spain, from an agreement with Norsk Hydro in 1912 to the first actual production in 1940. Bibliography.

341. JORDI GONZALEZ, R. "Peritajes industriales por dos farmacéuticos por emcargo de la Junta de Comercio de Cataluna, 1838-39," Boletin, Sociedad espanola de historia farmacia, 25 (1975), 48-66.
Describes the activities of T. Balvey and F. Muntada, for certain industrialists and the Chamber of Commerce of Catalonia, in the establishment of industrial products and monopolies, especially for the production of matches and the mineral acids.

342. ROCHE, Marcel. "Early History of Science in Spanish America," Science, 194 (1976), 806-09.
Finds the region's greatest strength in application, particularly in metallurgy and medical botany. Citations.

SWITZERLAND

HISTORIES

343. CIBA. The Story of Chemical Industry in Basel. Lausanne, 1959.
Although sponsored by one of the "fine chemical" manufacturers of Basel, it covers others. But it is no detailed history rather an elegantly illustrated coffee-table book.

344. CIBA. Beiträge zur Geschichte der Naturwissenschaft und der Technik in Basel. Olten, 1959.
A collection of articles, on the 75th anniversary of CIBA.

345. JACQUET, N. Die Entwicklung und volkwirtschaftliche Bedeutung der schweizerischen Teerfarbenindustrie. Basel, 1922.
Not seen.

346. JAUBERT, G.F. Historique de l'industrie Suisse des matières colorantes artificièlles. Geneva, 1896. (105 pp.)
A useful summary, with much reference to dyes and firms, with dates. No references.

UNITED STATES AND CANADA

EARLY WORKS

347. UNITED STATES. Census for 1820: Digest of Accounts of Manufacturers. Washington, 1823.
Reports on manufactures, including chemical, based on questionnaires. Financial and other data. An alphabetical list of products includes alum, "ashes," pearl and potash, brimstone (sulphur), gunpowder, lime, salt, saltpeter and vitriol.

348. UNITED STATES. Documents Relative to the Manufactures of the United States. Washington, 1833. (2 vols. House of Representatives, Documents, 22nd Congress, 2nd session)
Responses to a questionnaire distributed in 1831 include information on the production of salt, acids, lime, vitriol, and the consumption of chemicals in dye-making and the textile industry.

HISTORIES

349. "History of Chemical Engineering," Chemical and Metallurgical Engineering, 42 (1935), 177-228.
A series of short articles dealing with the United States. They are mostly trivial although those on unit operations and equipment (199-228) are of some value, being written by emminent engineers.

350. ABRAHAMS, Harold J. "Thomas Jefferson's Library of Applied Chemistry," Journal of the Elisha Mitchell Scientific Society, 77 (1961), 267-74.
Lists 21 books on "applied chemistry," cooking, wine-making, and brewing (3 each), bread, distilling and potash-making (2 each), and one each on confectionary, cider, maple sugar, dyeing, tar, and salt. Jefferson's library was to become the nucleus of the present Library of Congress.

351. AMERICAN CHEMICAL SOCIETY. DIVISION OF DYE CHEMISTRY, "Dye symposium," I&EC, 17 (1924) 409-19.
Articles by V.G. Bloede, Ellwood Hendrick, G.A. Prochazka, and Charles A. Monroe, the first three interesting reminiscences by participants in nineteenth

century attempts to make synthetic dyes in the United States.

352. ASH, Charles S., et al. "Symposium on the Contributions of the Chemist to American Industries," I&EC, 7 (1915), 273-304.
Numerous short pieces, some on unusual topics (asphalt, cottonseed oil). The objective seems to be to encourage the chemist (and the country) to counter the German industry. Says there are ten thousand chemists in the United States. No citations.

353. BAXTER, James P. III. Scientists against Time. Cambridge, Mass., 1946. (473 pp.)
Deals with scientific and technological innovations made in the United States during the Second World War. The principal chemical topics treated are explosives, pyrotechnics, penicillium, insecticides, war gases ("to deter the enemy from gas attacks"), the atomic bomb. Informative but dry. The author was "official historian" of OSRD (Office of Scientific Research and Development). No citations.

354. BENJAMIN, Marcus. "Some American Contributions to Technical Chemistry," Science, 21 (1905), 873-84.
Dubious history, with useful citations.

355. BISHOP, J. Leander. History of American Manufactures. Philadelphia, 1866. (2 vols)
Miscellaneous reports on the production of salt and saltpeter—and a single Prussian blue factory. Vitriol, alum, and soda are not mentioned (although there was in fact a vitriol works in Vermont).

356. BROWNE, C.A. "Chronological table of some leading events in the history of chemistry in America, from the earliest colonial settlements until the outbreak of the World War." I&EC, 18 (1926), 884-92.
Lists about 500 "events," from Medina's cold amalgamation process (1557) to the beginning of commercial production of radium in 1913, "hastily prepared partly from random notes which have been gathered by the author during the past ten years."

357. CARPENTER, Charles. "Coming of the Chemical Industry to Middle Appalachia," West Virginia History, 30 (1968-69), 535-47.
A circumstantial but rather unsystematic collection of notes, chiefly dealing with the Kanawha valley in the 1920's.

358. Clark, Victor S. History of Manufactures in the United States. Washington, 1916-29. (2 vols)
From 1607 to 1914. Includes leather, rubber, cement, pottery, glass, brewing, salt, sugar and fuels but mostly from the point of view of economic history. Citations.

359. CRANE, J.E. "Development of the Synthetic Ammonia Industry in the United States," I&EC, 22 (1930), 195-99.
A circumstantial account but undocumented and not very historical. Dates the first American factory "about 1921."

360. CRAWFORD, E.T., Jr. "Salt-pioneer Chemical Industry of the Kanawha Valley," I&EC, 27 (1935) 1109, 1274, 1411.
Fairly important especially for the 20th century.

361. DUPREE, A.H. Science in the Federal Government. Cambridge, Mass., 1957. (460 pp.)
A history of science (with technological implications) in the United States to 1940 in its relationship to the government. The important involvement of chemistry relates to activities of the Bureau of Chemistry of the Dept. of Agriculture. Citations.

362. GROSSMANN, Hermann. Die chemische Industrie in den Vereinigten Staaten, und die deutsche Handelsbeziehungen. Leipzig, 1912. (85 pp.)
Based on official documents. Begins with the situation as indicated by the census of 1860 and concludes with the tariff of 1912 and an analysis of the United States patent system in regard to chemicals.

363. GUEDON, J-C. Chemical Engineering by Design: The Emergence of Unit Operations in the United States.

Montreal: Univ. of Montreal, Inst. d'Histoire et de Sociopolitique des Sciences, 1979.
Not seen.

364. HALE, Harrison. American Chemistry: A Record of Achievement: The Basis for Future Progress. New York, 1921 (215 pp.)
Actually on industrial chemistry, water and sewage, food, textiles, coal tar dyes, etc., and not much concerned with history. But it has useful bibliographies.

365. HAYNES, Williams, and L. GORDY. Chemical Industry's Contribution to the Nation, 1635-1935. New York 1935. (a Supplement to Chemical Industries)
A series of disconnected articles; few of much value.

366. HAYNES, Williams. American Chemical Industry. New York, 1945-54. (6 vols)
A history. Vol. 1, to 1912, is especially well done, with many citations. Subsequent volumes cover shorter periods, with increasing attention to companies and managers until vol. 6, which as the subtitle, "Company histories."

367. HENAHAN, John F. "200 Years of American Chemicals," Chemical Week, 118 (1976) Feb. 25-40, 45-60.
Not seen.

368. HOUGEN, Olaf A. "Seven Decades of Chemical Engineering," Chemical Engineering Progress, 73 (1977) Jan.: 89-104.
In the United States, "from its humble beginnings" at the Massachusetts Institute of Technology in 1888. But it deals principally with the University of Wisconsin, where the author was a well-known teacher. Gives as "principal developments," industrial chemistry, for the decade 1906-15, unit operations, for 1916-25, material and energy balances for 1926-35, thermodynamics and process control for 1936-45, applied kinetics and process design for 1946-55, and transport phenomena, process dynamics, process engineering and computer technology for 1956-65. Indicates that from about 1955 undergraduate instruction contained more chemical engineering than chemistry. No references but several interesting graphs.

369. IVEY, Dean B. "Origins of the American Synthetic Dye Industry, 1865-1925, with special emphasis upon government policy." Dissertation (MA), Univ. of Delaware, 1963.
An intelligent analysis of a crucial aspect of the history of the American chemical industry. Citations.

370. JOHNSON, Howard C.E. "The Rise of the U.S. Chemical Industry, 1918-68," Chemical Week, 103 (1968) Nov., 104-45.
A capsule history without references, with many illustrations (portraits of Lindberg, Al Capone, and others of equal relevance). Perhaps useful to those totally ignorant of the subject.

371. JONES, Daniel P. "The Role of Chemists in Research on War Gases in the United States during World War I." Dissertation, Univ. of Wisconsin, 1969.
Not seen.

372. KIRKPATRICK, S.D., ed. Twenty-five Years of Chemical Engineering Progress. New York: American Inst. of Chemical Engineers, 1933. (373 pp.)
Twenty-five articles, by as many chemical engineers, who pretty much cover the field. Only seven of them have citations although others give statistics. A useful volume although the general tone is indicated by the appearance of the words "progress," or "advance" in ten titles.

373. KIRKPATRICK, S.D. and J.R. CALLAHAN, eds. "Builders of the Chemical Century," Chemical Engineering, 59 (1952), 145-210.
The 50th anniversary of the American Institute of Chemical Engineers is celebrated in a series of brief articles, mostly in journalese with minimal historical content. Portraits of a number of leaders.

374. KRAMMER, Arnold. "Technology Transfer as War Booty: the United States Technical Oil Mission to Europe, 1945," T&C, 22 (1951), 68-103.
"Germany's substantial advances in numerous industries --especially rocketry, optics, plastics, industrial chemistry, pharmaceuticals, and synthetic fuel technology--would be of substantial value both to American

industry and to the continuing war effort against the Japanese." Here exemplified by synthetic petroleum. A thought-provoking account of American zeal for German technology and what became of it. Citations.

375. MAHONEY, Tom. *The Merchants of Life: An Account of the American Pharmaceutical Industry.* New York, 1959. (278 pp.)
Popular, without references. Consists principally of chapters devoted to the major companies.

376. MICHAEL, T.H.G. "The Association of Aspects of Chemical Engineering in Canada," in *Furter* (1980, no. 116) 199-204.
Discusses professional organizations, regulatory bodies, etc. No references.

377. NOYES, Wm. A., Jr., ed. *Chemistry: A History of the Chemistry Components of the National Defense Research Committee.* Boston, 1948. (524 pp.)
A fact-crammed but little documented history of wartime research in explosives, gas warfare, aerosols, and fuels. A fullsome use of acronyms and a determination to mention everybody involved make it a chore to read.

378. SADTLER, S.P. "Early Chemical Manufacture in Philadelphia," *I&EC*, 8 (1916), 1153-56.
Summary to the end of the nineteenth century.

379. SERVOS, John W. "The Industrial Relations of Science: Chemical Engineering at MIT, 1900-1939," *Isis*, 71 (1980), 531-49.
On an experiment to determine whether "industrial sponsorship alone (can) provide a stable and vital basis for applied scientific research in an academic setting." Citations, including faculty correspondence.

380. SHEMILT, L.W. "A Century of Chemical Engineering Education in Canada," in *Furter* (1980, no. 116), 167-98.
Narrative history, the first highlight being the appointment of a professor of applied chemistry at Toronto in 1882. Gives details on the locations of schools, degrees awarded, research activity "indicators." Citations.

381. STEIN, Charles M.A. "The Rise of the Organic Chemical Industry in the United States," I&EC, 32 (1940) 137-44. (reprinted in SIAR [1940], 177-92)
An upbeat survey of "phenomenal growth" since 1914, being remarks on the occasion of the award of the Perkin medal to the speaker.

382. TURRILL, P.L. "Studies in the Mineral and Chemical Resources of the Mojave Desert," JCE, 9 (1932) 1319-39, 1531-52, 2041-64.
A description of this California mineral cornucopia, with a strong historical orientation.

383. TUTTLE, Wm. M., Jr. "The Birth of an Industry: The Synthetic Rubber 'Mess' in World War II," T&C. 22 (1981), 35-67.
History of the United States synthetic rubber program, mostly political but implying that the practitioners of science and technology deserve most of the credit. He doesn't prove it, but the rubber was in fact made.

384. UNITED STATES. DEPT. OF COMMERCE. BUREAU OF FOREIGN AND DOMESTIC COMMERCE. The American Chemical Industry. Production and foreign trade in the first quarter of the twentieth century. Washington, 1929. (Trade Promotion Series, no. 78)
Mostly statistical.

385. UNITED STATES. BUREAU OF THE CENSUS. Historical statistics of the United States, 1789-1945. Washington: GPO, 1949.
Includes data on the production of salt, soda ash (artificial soda) and sulphuric acid.

386. VAN ANTWERPEN, F.J. and Sylvia FOURDRINER, eds. First Fifty Years of the American Institute of Chemical Engineers. New York, 1958. (188 pp.)
Concerned only with the history of the society.

387. WARRINGTON, C.J.S., and R.V.V. NICHOLLS. A History of Chemistry in Canada. New York, 1949. (502 pp.)
A narrative without citations chiefly devoted to industrial chemistry. Bibliography.

388. WARRINGTON, C.J.S. and B.T. NEWBOLD. Chemical Canada, Past and Present. Ottawa: Chemical Institute of Canada, 1970. (290 pp.)
Unlike the history (no. 387) this is "chronological rather than topical." Most useful for its actual chronology, the text being telegraphic and with citations.

389. WEBER, H.C. "The Improbable Achievement: Chemical Engineering at MIT," in Furter (1980, no. 116), 77-96.
A straightforward narrative based on communications. No references. It doesn't tell what was improbable about it.

390. WESTWATER, J.W. "The Beginnings of Chemical Engineering Education in the U.S.A.," in Furter (1980, no. 116), 141-52.
Ruminations about the genesis of chemical engineering education, largely in terms of the appearance of the name. Numerous references to college catalogues and departmental histories.

391. WHITE, Alfred H. "Chemical Engineering Education in the United States," Trans., AIChE, 21 (1928), 55-85.
The chemical engineer emerged in the U.S. Census in 1910, and the number of students trebled in the following decade. Other miscellaneous information, e.g., the chemical engineering curricula in 1925 averaged 13.9% "cultural" subjects and only 10.3% "chemical engineering." References.

392. WIK, Leonard. "Henry Ford's Science and Technology for Rural America," T&C, 3 (1962), 247-58.
The application of chemistry to agriculture and to the exploitation of agricultural products was part of Ford's solution to the "farm problem" of the 1920's, and culminated in the Farm Chemurgic Council of 1935. Citations.

393. WILBERT, M.I. "Early Chemical Manufactures. A Contribution to the History of the Rise and Development of Chemical Industries in America," JFI, 157 (1904), 365-78.

Related primarily to Philadelphia and based on secondary works. But some are not readily available, such as Henry Simpson's <u>Lives of Emminent Philadelphias Now Deceased</u> (1859).

394. WILKINSON, Norman. "Brandywine Borrowings from European Technology," <u>T&C</u>, 4 (1963), 1-13.
Discusses early nineteenth century European influence on American milling, tanning, gunpowder and papermaking. Citations.

BIOGRAPHIES

COLLECTIONS

395. "Pioneers of Electrochemistry," <u>Electrochemical Industries</u>, 1 (1902)
This journal began with a flourish of interest in history, with very brief biographies of C.M. Hall (10), A.H. Cowles (56), E.G. Acheson (90-91), H.Y. Castner (121), Wm. Ostwald (214-15), and C.S. Bradley (451-52).

396. ALLEN, J. Fenwick. <u>Some Founders of the Chemical Industry</u>. London, 1906. (289 pp.)
Biographies of William Gossage, J.C. Gamble, James Muspratt, Andreas Kurtz, Henry Deacon, James Shanks, Christian Allhusen, and Peter Spence, "small manufacturers" whose breed, according to the author, has been saved from extinction by the formation in 1890 of the United Alkali Company. These biographies had originally been published in the <u>Chemical Trade Journal</u>.

397. BUGGE, Günther, ed. <u>Das Buch der grossen Chemiker</u>. Berlin, 1929-30. (2 vols)
Particularly relevant to technology are V. Biringuccio, Georg Agricola, J.R. Glauber, N. Leblanc, A.W. von Hofmann, Peter Griess, H. Caro, Adolph Frank, Clemens Winkler, George Lunge, and Heinrich von Brunck.

398. CHEMICAL SOCIETY (London). <u>Memorial Lectures</u>. London, 1901-42. (4 vols)
Includes some chemists relevant to the industry, A.W. von Hofmann, Henri Moissan, Adolf von Beyer, and Fritz Haber.

399. FARBER, Eduard, ed. Great Chemists. New York, 1961. (1,642 pp.)
Includes Anselm Payen, A.W. von Hofmann, Ernest Solvay, and Wm. Perkin (the latter by S.M. Edelstein, and reprinted from American Dyestuff Reporter, 45 (1956), 598-608).

400. FISCHER, Ernst. "Die Gründer der Farbewerk Hoechst," Nassauische Lebensbilder, 6 (1951), 252-61.
Eugen Lucius, Wilhelm Meister, Adolf Brüning.

401. HAYNES, Williams. Chemical Pioneers: The Founders of the American Chemical Industry. New York, 1939. (288 pp.)
The pioneers, each the subject of a short undocumented biography, are John Winthrop, Jr., G.D. Rosengarten, Martin Kalbfleisch, Alexander Cochrane, J.J. Mapes, E.R. Graselli, George T. Lewis, Lucien C. Warner, Edward Mallinckrodt, August Klipstein, E.C. Klipstein, Martin Dennis, Jacob Hasslacher, J.F. Queeny, Frank S. Washburn, H.H. Dow.

402. MILES, Wyndham D., ed. American Chemists and Chemical Engineers. Washington: American Chemical Society, 1976. (544 pp.)
507 biographies by various authors, the editor himself being the most substantial contributor. Valuable.

403. OBERDORFFER, Kurt, ed. Ludwigshafener Chemiker. Düsseldorf, 1958. (2 vols)
Carl Bosch, Heinrich von Brunck, Heinrich Caro, Georg Giulini, Carl Grünzweig, Rudolf Knietsch, Albert Knoll, Alwin Mittasch, Fritz Raschig, Ludwig Reimann (Jr. and Sr.), Fritz Winkler.

404. WILLIAMS, Glenn C., and J.E. VIVIAN. "Pioneers in Chemical Engineering at MIT," in Furter (1980, no. 116), 113-28.
Capsule biographies of faculty members at the Massachusetts Institute of Technology. No references.

INDIVIDUAL BIOGRAPHIES (alphabetical by subject)

405. ACCUM, Frederick (1769-1838). Browne, C.A., "The Life and Chemical Services of Frederick Accum," JCE, 2 (1925), 829-51, 1008-34, 1140-49.

406. ACHESON, E.G. (1856-1931). Szymanowitz, Raymond. Edward Goodrich Acheson. New York, 1971. (628 pp.) Acheson, who worked briefly for Edison, is one of the more successful "graduates" of the Menlo Park laboratory, being known for his commercial development of "carborundum" and artificial graphite. This admiring biography by his long-time associate, is more valuable than most because of its extensive use of correspondence, diaries, and family records. References.

407. BAEKELAND, Leo H. (1863-1944). Matthis, A.R. Leo H. Baekeland. Brussels, 1948. (73 pp.)
An extended éloge. No citations.

408. BAEKELAND, Leo H. Kaufmann, Carl B. Grand Duke, Wizard and Bohemian, A Biographical Profile of Leo H. Baekeland. 1968. Typescript in the Library of Congress. (139 pp.)
Citations.

409. BAIST, Ludwig (1825-99). Fleming, H.W. Ludwig Baist, der Gründer der chemischen Fabrik Griesheim. Munich, 1965. (83 pp. Tradition, Beiheft 4)
Based on documents of the firm.

410. BERGIUS, Friedrich (1884-1945). Schmidt-Pauli, E. von. Freidrich Bergius: ein deutscher Erfinder kampft gegen die englische Blockade. Berlin, 1943.
Undocumented but informative. The title and certain other peculiarities are explained by the date of publication.

411. BOSCH, Carl (1874-1940). Holdermann, Karl. Im Banne der Chemie: Carl Bosch, Leben und Werk. Düsseldorf, 1948. (328 pp.)
Few citations. Indicates that unpublished material has been used.

412. BOSCH, Carl. Holdermann, Karl. Carl Bosch. Düsseldorf, 1955.
No citations.

413. BRUNCK, Heinrich von (1847-1911). Glaser, C., Heinrich von Brunck," Berichte DCG, 46 pt. 1 (1913), 353-80.
Important.

414. CARO, Heinrich (1834-1910). "From Alchemist's Furnace to BASF: Heinrich Caro's Years of Training," Die BASF, 24 (1974), 3-10.
Too brief to be very useful and without references.

415. CASTNER, Hamilton Y. (1858-99). Lord, V.H., "Hamilton Young Castner," JCE, 19 (1942), 353-56.

416. COTTRELL, F.G. (1877-1948). Cameron, F.T. Cottrell, Samaritan of Science. New York, 1952. (414 pp.)
Cottrell was notable as an academician, entrepreneur, and philanthropist. No indication of sources.

417. DOW, H.H. (1866-1930). Campbell, M., & H. Hatton. H.H. Dow. New York, 1951. (186 pp.)
An "authorized" biography but useful. Based on company archives without references.

418. DUISBERG, Carl. (1861-1935). Abhandlungen, Vorträge, und Reden aus den Jahren 1882-1921. Berlin, 1923. Abhandlungen, etc., 1923-33. Berlin, 1933. (2 vols)
Reprints documents (e.g., patents), articles, speeches, and other material relative to the Verein Deutsche Chemiker, Deutsches Museum, the firm of Bayer, and his personal work. There are a few historical pieces, e.g., "Die Einfluss Liebigs auf die Entwicklung der Chemische Industrie" (which was also published in English in Popular Science, April 1904.

419. DUISBERG, Carl. Meine Lebenserringerungen. Leipzig, 1933. (207 pp.)
No citations.

420. DUISBERG, Carl. Flechtner, H.J. Carl Duisberg. Düsseldorf, 1949. (413 pp.)
Indicates use of Duisberg's unpublished materials from the firm of Bayer, but the citations are to published materials.

421. DUNDONALD, Lord (Archibald Cochrane, 1749-1831). Clow, Archibald, and Nan Clow, "Lord Dundonald," Economic History Review, 13 (1942), 47-58.
A biographical sketch, which also aims to show that the introduction of coal gas was a consequence of a search for uses for coal by-products through the last half of the eighteenth century. Citations.

422. ENGLEHORN, Friedrich (1821-1902). Jacob, F. "Friedrich Engelhorn," <u>Schriften der Gesellschaft der Freunde Mannheims</u> (Mannheim), 8 (1959).
Not seen. Engelhorn was a founder (1865) of BASF.

423. GARBETT, Samuel (1717-1803). Bebbington, P.S. <u>Samuel Garbett, a Birmingham pioneer</u>. Dissertation, University of Birmingham, 1938.
Reports that Garbett hit upon a new method of making sulphuric acid (the chamber process) "almost immediately" after 1746, when he and John Roebuck had set up a laboratory for gold-refining.

424. GLAUBER, J.R. (1603-68). Pietsch, Erich. <u>Johann Rudolph Glauber, der Mensch, sein Werk und seine Zeit</u>. Munich, 1956. (64 pp. Deutsches Museum, Abhandlungen und Berichte, 24, Heft 1)
Citations.

425. GLAUBER, J.R. Spronsen, J.W. van, "Glauber grondlegger van chemische industrie." <u>Nederlandse Chemische Industrie</u>, 5 (1970), 3-11.
References.

426. GREISS, Peter. (1829-88). Hofmann, A.W. von and Emil Fischer, "Zur Erinnerung an Peter Griess," <u>Berichte</u>, DCG, 24, Th. 1 (1891), 1007-78.
Important for its inclusion of a detailed discussion of Griess' work with azo dyes. It is followed by a 38 pp. supplement by H. Caro.

427. GREISS, Peter. Ward, E.R., "Peter Griess (1829-1888) and the Burton Breweries," <u>Journal, Royal Institute of Chemistry</u>, 82 (1958), 383-89.
A well-written account of Griess' years as a "brewers chemist" and of science at Burton-on-Trent in his time. Citations.

428. GRIESS, Peter. Bolton, John, "The Life and Times of Peter Griess," <u>Journal, Society of Dyers and Coulorists</u>, 75 (1959), 278-85.
Summary. No citations, but incorporates material from letters presented by the Griess family to ICI. Illustrated.

429. HAZARD, Rowland (1829-1898). Haynes, Williams, "Rowland Hazard, the Father of the American Alkali Industry," Chemical Industries, 47 (1940), 248-53.
Worshipful and without citations. Hazard introduced the Solvay process in the United States.

430. HOFMANN, A.W. von (1818-92). Perkin, W.H., et al. "Hofmann Memorial Lecture(s)," Trans, Chemical Society (London), 69 (1896), 575-732.
Extensive recollections of the doyen of dye chemists. Perkin's own piece is on "The Origin of the Coal Tar Industry and the Contribution of Hofmann and His Pupils."

431. IPATIEFF, V.N. (1867-1952). The Life of a Chemist. Stanford, Calif., 1946. (658 pp.)
His memoires. Ipatieff is famous for his commercial applications in high-pressure catalytic synthesis in Russia and the United States.

432. KEIR, James (1735-1820). Moilliet, Amelia. Sketch of of the Life of James Keir. London, 1871. (164 pp.)
Keir established an artificial soda factory at Tipton, near Dudley, about 1780.

433. LEBLANC, Nicolas (1753-1806). Anastasi, A. Nicolas Leblanc: sa vie, ses traveaux et l'histoire de la soude artificielle. Paris, 1884.
By Leblanc's grandson. Its value is discussed in Gillispie, 1957 (no. 1492).

434. LEBON, Philippe (1769-1804). Fayol, Amédée. Philippe Lebon, inventeur du gaz d'éclairage. Paris, 1943. (90 pp.)
No citations but some unpublished sources are listed in the brief bibliography.

435. LEWIS, Warren K. (1882-1975). Lewis, H.C., "W.K. Lewis," in Furter, (1980, no. 116), 129-40.
Lewis was first head of chemical engineering at MIT and is often called "the father" of the profession. This chatty sketch of personal idiosyncrasies tells us, among other things, that Lewis was convinced "that history is essential...for one thing most engineering students have little interest in problems of the past,

but for future use they need some kind of feel for how the greatest scientific and engineering minds work when they run into perplexing problems." No citations.

436. LUNGE, Georg (1839-1923). Berl, E. "Georg Lunge," JCE, 16 (1939), 453-60.
No references.

437. MOND, Ludwig (1839-1909). Armstrong, H.E., "The Monds and Chemical Industry, a Study in Heredity," Nature, 127 (1931), 238-40.
Ruminations on the careers of Ludwig and Alfred Mond, aimed to show "that biography should be the recognized province of the structural chemist." (sic) No citations.

438. MOND, Ludwig. Cohen, J.M. The Life of Ludwig Mond. London, 1956. (295 pp.)
Based on interviews with descendents and associates.

439. MUSPRATT, James (1793-1886). Stephens, M.D., and G.W. Roderick, "The Muspratts of Liverpool," AS 29 (1972), 287-311.
James Muspratt (1793-1886), "father of the heavy chemical industry at Merseyside" (Liverpool), and two generations of his sons are dealt with, using family papers in the Liverpool Central Libraries.

440. NOBEL, Alfred (1833-96). Sohlman, Ragnar, and Henrik Sabück. Nobel. Dynamite and Peace. New York, 1929. (353 pp.)
The standard biography authorized by the Nobel-Stiftalsen, Stockholm. No citations, but appendices describe his patents and print an interview with the inventor concerning nitroglycerine and dynamite.

441. NOBEL, Alfred. Halasz, Nicholas. Nobel. A Biography of Alfred Nobel. New York, 1959. (281 pp.)
Popular and without citations although it is indicated that manuscript sources in Sweden and elsewhere were used.

442. NOBEL, Alfred. Bergengren, Erick. Alfred Nobel. London, 1962. (222 pp.)
From the Swedish edition of 1960.

443. PERKIN, W.H. (1838-1907). Robinson, Robert. "The Life and Work of Sir William Henry Perkin," in White, (1956, no. 1262), 41-50.
Very miscellaneous but contains some uncommon notes as on the origin of the name "mauve."

444. ROEBUCK, John (1718-95). Jardine, R. "An Account of John Roebuck," Trans., Royal Society of Edinburgh, 4 (1796), 65-87.

445. ROEBUCK, John. Clow, Archibald, and Nan Clow, "John Roebuck," Chemistry and Industry, 61 (1942), 497-98.
An important biographical sketch without citations.

446. SOLVAY, Ernest (1838-1922). Héger, P.F., and C. Lefébure. La vie d'Ernest Solvay. Brussels, 1929. (164 pp.)
Not seen.

447. SWINDIN, Norman (1880-1976). Engineering without Wheels: A Personal History. Letchworth, 1962. (255 pp.)
The candid autobiography of a British free-lance chemical engineer, whose peripatetic career touched most of the heavy chemical industry.

448. WINNACKER, Karl (1903-). Challenging Years, My Life in Chemistry. London: Sidgwick & Jackson, 1972. (440 pp.)
The author entered the employ of Hoechst in 1933, when it was part of IG Farbenindustrie, and survived to reach "top management" in the postwar Hoechst. This is less a story of chemical technology than successful management, and the postwar occupation appears to be more regretted than either the "third reich" or World War II. The German original (1971) was titled Nie den Mut verlieren.

COMPANY HISTORIES

These are elusive publications, very often commemorative rather than historical, often privately published, and occasionally not "published" at all (Haber, no. 122: pp. 232-35, mentions a number in the later category). The historical content is often very small indeed, but I've included some of these as the information is otherwise hard to come by. Coverage of traditional industries (rubber, glass, etc.) is limited to those which indicate some peculiarly chemical activity, and coverage is altogether limited to histories found in the Library of Congress, the Smithsonian Libraries, and the Baker Library of Harvard University.

COLLECTIVE

449. MARCUS, Alfred. <u>Die grossen Chemiekonzerne</u>. Leipzig, 1929.
Describes the establishment of the large concerns of the time, L'Air Liquid, Imperial Chemical Industries (ICI), IG Farbenindustrie, Kuhlmann, Du Pont, and Montecatini.

450. TRADITION. ZEITSCHRIFT FUR FIRMEN-GESCHICHTE UND UNTERNEHMERBIOGRAPHIE (1956 ff., since vol. 22 titled ZEITSCHRIFT FUR UNTERNEHMENS)
This important journal deals with all aspects of business history, but almost exclusively German and with little attention thus far to the chemical industry. Some articles are here entered separately. Includes annual bibliographies.

INDIVIDUAL (alphabetical by company)

451. AFRICAN EXPLOSIVES & CHEMICAL INDUSTRIES. Cartwright, Alan P. <u>The Dynamite Company</u>. Capetown, 1964. (267 pp.)
Published by the company. Journalistic but informative. No citations.

452. ATLAS CEMENT CO. Hadley, Earl J. <u>The Magic Powder: History of the Universal Atlas Cement Company and the Cement Industry</u>. New York, 1945. (382 pp.)
Informative but with little indication of sources.

453. BADISCHE ANILIN UND SODA-FABRIK (BASF). Steinert, O., and W. Roggersdorf. <u>In the Realm of Chemistry</u>. Düsseldorf, 1965. (155 pp. There was a simultaneous German edition)
Subtitled "pictures from the past and present of the Badische Anilin & Soda-Fabrik," and it is mainly pictures--a centenial publication. Those of the works at

Ludwigshafen in 1881 (sixteen years after the founding of the company), 1945 (after 127 air attacks), and 1965, are thought provoking. This was and may still be the largest chemical works in the world.

454. BASF. Ludwig, W. "Highlights in the history of BASF," Trans., Institute of Chemical Engineers, 44 (1966). Not seen.

455. BASF. Wolf, Gerhardt. Die BASF. Vom Werden eines Weltunternehmens. Ludwigshafen, 1970. (60 pp., SB no. 6)
A brief history of the company prepared for summer student employees.

456. BASF. Schuster, Curt. Vom Farbenhandel zur Farbenindustrie: Die Erste Fusion der BASF. Ludwigshafen, 1973. (87 pp., SB no. 11)
On the early history of the company.

457. BASF. Mach, Erich. Entwerfen und Bauen. Ludwigshafen, 1975. (137 pp. SB no. 13)
On the history of the firm with particular reference to the construction of works.

458. BASF. Schuster, Curt. Wissenschaft und Technik. Ihre Begegnung in der BASF während der ersten Jahrzehnte der Unternehmungsgeschichte. Ludwigshafen, 1976. (137 pp., SB no. 14)
Deals with the scientists associated with the firm during its first decade, Carl Graebe, Carl Liebermann, Adolf von Baeyer, Peter Griess, Emil Fischer, and Carl Engler.

459. BASF. Queisner, Rudolph, and Kurt Schliesser. Zirkel-Schraubstock-Elektronik. 50 Jahre BASF-Lehrwerkstätten. Ludwigshafen, 1977. (121 pp., SB no. 15)
On the history of training at BASF.

460. BAYER (Farbenfabriken Bayer Aktiengesellschaft). Pinnow, Hermann. Werkgeschichte: der Gefolgschaft der Werke Leverkusen, Elberfeld und Dormagen. Munich, 1938. (201 pp.)
Narrative account of the first 75 years (1863-1938) of Bayer (part of IG Farbenindustrie in 1938) without citations. Illustrated.

461. BAYER. Revolution im Unsichtbaren. Düsseldorf, 1963. Mostly pictures, over 150 unnumbered pages. Useful for a chronology of its history. No citations.

462. BRUNER, MOND. Dick, W.F.L. A Hundred Years of Alkali in Chesire. n.p.: ICI, 1973. (125 pp.) Founded in 1873 as Bruner & Mond, to manufacture soda by the Solvay process. Now a division of ICI.

463. CASTNER KELLNER. Williams, T.I. Fifty years of progress: the story of the Castner Kellner Alkali Co. 1895-1945. London: ICI, 1945. Not seen.

464. CHANCE. Chance, J.F. A history of the firm of Chance Bros. and Co., glass and alkali manufacturers. London, 1919-26. (310 pp.) Not seen.

465. CHANCE. 100 years of British glass-making, 1824-1924. n.p., n.d. (23 pp.) An informative summary history of Chance Bros. without references.

466. CIBA. Gesellschaft für chemische Industrie in Basel, 1884-1934. n.p., n.d. (91 pp.) An account of "prehistory" (1864-84) is followed by a good summary history of the firm without citations. Illustrated.

467. COMMERCIAL SOLVENTS. Kelly, F.C. One Thing Leads to Another. Boston, 1936. (104 pp.) A chatty popular history of the company founded in 1920. No references.

468. CONSOLIDIRTE ALKALIWERKE WESTEREGLN. Fünfzig Jahre Consolidirte Alkaliwerke Westeregln, n.p., 1931. (324 pp.) Founded in 1881, its principal asset being the Kali works at Douglashall, the firm consequently produced Braunkohl and chemicals (calcium chloride, bromine, cyanides, electrochemicals). This serious and well-written history, illustrated from company records, is a model of its kind.

469. CROSFIELD. Musson, A.E. Enterprises in Soap and Chemicals. Joseph Crosfield & Sons Ltd., 1815-1965. Manchester, 1957.
Still primarily a soap-maker, Crosfield was absorbed in 1919 by Lever (from 1929 Unilever).

470. DEUTSCHEN SOLVAY-WERKE. 75 Jahre deutsche Solvay-Werke, 1880-1955. Darmstadt, 1955.
Not seen.

471. DOW. Whitehead, Don. The Dow Story: the History of the Dow Chemical Company. New York, 1968. (298 pp.)
Popular account without citations. The company was founded in the 1890's.

472. DU PONT, E.I. DE NEMOURS. Rideal, C.F., and A.W. Atwood. The History of the E.I. Du Point de Nemours Powder Co. New York, 1912. (224 pp.)
Published by Business America, otherwise known as Banker and Investor Magazine, and subtitled "A Century of Success," this is a quaint effusion of Victorian business history but not without information.

473. DU PONT. Dutton, W.S. Du Pont, One Hundred and Forty Years. New York, 1942. (406 pp.)
A comprehensive history of "the Du Pont Company as seen by Du Pont men...an 'inside' view." No references beyond notice that company and family records have been utilized.

474. DU PONT. Senecal, V.E. "Du Pont and Chemical Engineering in the 20th Century," in Furter (1980, no. 116), 283-302.
Summary. Du Pont's Engineering Dept. was organized in 1902 and its Experiment Station in 1903. References.

475. ELEKTROCHEMISK a/g. Peterson, Erling. Elektrochemisk a/g. Oslo, 1953. (252 pp.)
No references.

476. EXXON. Gornowski, E.J. "The History of Chemical Engineering at Exxon," in Furter (1980, no. 116), 303-12.
Standard Oil was organized in 1882, Jersey Standard in 1911. The Development Dept. was organized in 1919. Describes some of its work.

477. FIRESTONE TIRE & RUBBER CO. Lief, Alfred. The Firestone Story. New York, 1951. (437 pp.)
History of the company, which was founded in 1900. No citations but bibliographic reference to "company source material."

478. GAS LIGHT & COKE CO. Everard, S. The History of Gas Light and Coke Company, 1812-1949. London, 1949. (428 pp.)
Well written and authoritative from company records but without citations.

479. GEIGY AG. Bürgin, Alfred. Geschichte des Geigy Unternehmens von 1758 bis 1929. Basel, 1952. (325 pp.)
One of the best histories of a firm, dealing with matters economic, technical and social with ample citation. Continued in Geige Heute, Basel, 1958 (288 pp.). Yet another book, 200 Jahre Geigy (Basel, 1958), is a summary account in which history is incidental.

480. GENERAL CHEMICAL CO. The General Chemical Company after Twenty Years, 1899-1919. New York, 1919. (103 pp.)
A summary, without citations. General Chemical was an amalgamation of pre-existing firms.

481. GOLDSCHMIDT, AG. (Däbritz, Walther) Th. Goldschmidt AG. Geschichte einer deutschen chemischen Fabrik. Essen, 1937. (131 pp.)
90th anniversary history of a manufacturer of textile chemicals. No references.

482. GOODYEAR TIRE AND RUBBER CO. Allen, Hugh. The House of Goodyear. 2nd ed., Cleveland, 1949. (591 pp.)
Revision of a company-sponsored history first published in 1936. The company was founded in 1898. No references.

483. GOODYEAR TIRE AND RUBBER CO. Dietz, David. Harvest of Research: The Story of the Goodyear Chemical Division. Akron, 1955.
Not seen.

484. GRIESHEIM-ELEKTRON. Raschen, H., and P. Hoffmann. *75 Jahre Chemische Fabrik Griesheim-Elektron.* Griesheim, 1938. (128 pp.)
Not seen. The firm was founded at Frankfurt/M in 1856, as Chemische Fabrik Griesheim.

485. GRIESHEIM-ELEKTRON. Pistor, Gustav. *Hundert Jahre Griesheim, 1856-1956.* Tegernsee, 1958. (245 pp.)
Centennial history of a manufacturer of agricultural chemicals, which added electrolytic alkali and chlorine in 1890.

486. GRIESHEIM. Forstmann, Wilfried, "Die chemische Fabrik Griesheim in der Grossen Depression," *Tradition*, 26 (1981), 42-60.
Using records of Griesheim, he tests Hans Rosenberg's theory of a "Great Depression," 1873-94, and finds it wanting. Citations. Tables.

487. HARBISON WALKER REFRACTORIES CO. MacCloskey, Jas. E., Jr. *History of Harbison Walker Refractories Co.*, Pittsburgh, 1952. (134 pp.)
Mostly business history. The author joined the company shortly after its foundation in 1902 out of the Star Fire Brick Co., which dated from 1865. No references.

488. HENKEL. *Hundert Jahre Henkel.* Düsseldorf, 1976. (201 pp.)
A picture book with minimal text. The firm makes cleansing materials.

489. HOECHST. Baumler, Ernst, ed. *Ein Jahrhundret Chemie.* Düsseldorf, 1963.
Deals with the technical and financal history of the firm, based on internal documents but without citations. Illustrated.

490. HOECHST. *Der Hoechst-Konzern Entsteht.* Die Verhandlungen über die Auflösung von IG Farben und die Gründung der Farbwerk Hoechst AG. 1945-1953. Frankfurt/M, 1978. (236 pp. DHA nos. 49 and 50)
Documents on the resurrection of Hoechst after the Second World War.

491. HOOKER. Thomas, Robt. E. Salt & Water, Power & People. A Short History of Hooker Electrochemical Co., Niagara Falls, N.Y., 1955 (109 pp.)
A summary, without references.

492. HULS. Kränzlein, Paul. Chemie im Revier. Düsseldorf: Econ, 1980. (366 pp.)
Narrative history based on company archives, but without citations. The company was founded by I.G. Farbenindustrie in 1938 to make synthetic rubber and become independent after 1945.

493. ICI (Imperial Chemical Industries). Reader, W.J. History of ICI: The First Quarter Century, 1926-1952. Oxford: Oxford Univ. Press, 1970-75. (2 vols.)
A serious history with citations and not as limited as it sounds. The first 450 pages of vol. 1 are occupied with the history before 1926, that is, with the predecessor companies.

494. IG FARBENINDUSTRIE. Sasoly, Reinhard. IG Farben. New York, 1947. (310 pp.)
Journalistic.

495. IG FARBENINDUSTRIE. Ter Meer, Fritz. Die IG Farben-Industrie Aktiengesellschaft. Düsseldorf, 1953. (115 pp.)
An attempt to recover the history of IG Farben from "the political controversies" surrounding it. Concludes that its principal significance was in the realm of science, technology and social conditions. No references or index. The author was a top officer of the company.

496. IG FARBENINDUSTRIE. Petzina, Dieter, "IG Farben und Nationalsozialistisches Autarkiepolitik," Tradition, 13 (1968), 250-54.
"An economic autarchy lay in the interest of both (the Nazi government and IG Farben) after 1933. Citations.

497. KALLE. Voelcker, Heinrich. 75 Jahre Kalle, 1863-1938. Wiesbaden-Biebrich, n.d. (1938 ?). (207 pp.)
Kalle was an early dye firm. Bibliography.

498. KERR-McGEE. Ezell, John S. Innovations in Industry: the Story of Kerr-McGee. Norman, Okla.: Univ. of Oklahoma Press, 1979. (542 pp.)
The firm existed from 1932, as A&K Petroleum Co., primarily for drilling, from which it expanded into uranium (1952), potash, and other minerals, all of which are discussed here. Citations.

499. KOSTA. Anderbjork, J.E. Kosta Glasbruk, 1742-1942. Jubileumskrift. Stockholm, 1942. (231 pp.)
Gives substantial attention to technology. Bibliography and reference to archives used.

500. KUHLMANN. Cent ans d'industrie chimique: les Etablissements Kuhlmann, 1825-1925. Paris, 1926 (137 pp.)
Not seen.

501. KUNHEIM. Spiegel, L. 100 Jahre Kunheim. Berlin. 1925. (33 pp.)
The firm was founded in 1826, to make soda, potash, sal amoniac, pyroligneous acid, animal charcoal, white lead and sugar of lead. No citations.

502. MATHIESON. Fifty Years of Chemical Progress. New York, 1942. (46 pp.)
A history of Mathieson Alkali Works published by the company. Very cursory but what there is of it is on the history of the company, not (as frequently) on biographies and futuristic speculation.

503. MATTHES & WEBER. (Däbritz, Walther). E. Matthes & Weber AG, Duisburg. Duisburg, 1938. (208 pp.)
A centenary history, 1838-1938, of a firm based on a sulphuric acid works (then 14 years old) to which artificial soda was added in 1838, being the fourth Leblanc works in Germany. No references.

504. MEISTER, LUCIUS und BRUENING. Farbwerke vorm. Meister, Lucius & Brühning, 1863-1913. Hoechst, n.d. (55 pp.)
An historical sketch with some economic data and a description of the Hoechst works with plans and sketches. No references.

505. MEISTER, LUCIUS and BRUENING. Pinnow, Hermann. Zur Erinnerung an die 75 Wiederkehr des Gründungstages der Farbewerke vormals Meister, Lucius und Brüning. Munich, 1938. (192 pp.)
A straightforward account of the firm which began as Meister, Lucius & Brüning, evolved into Hoechst and was in 1938 part of IG Farben. Informative although written in the fervent patriotic style typical of the time and place of publication. No references.

506. MERCK. Merck, 1668-1968. Darmstadt, 1968. (about 150 unnumbered pages)
A picture book with minimal text. It tells us that the American firm of the same name was founded in 1891 and has no connection with the German Merck. No references.

507. MICHIGAN ALKALI CO. Pound, Arthur. Salt of the earth. the story of Captain J.B. Ford and Michigan Alkali Co. 1890-1940. Boston, 1940. (122 pp.)
Mostly a family history of Ford (1811-1903), the "founder of American plate glass." He made plate glass from the 1860's, one of his factories becoming Pittsburgh Plate Glass Co. He also founded Michigan Alkali Co. in 1891, when he was 80. A few citations.

508. MONSANTO. Forrestal, D.J. Faith, Hope, and $5000: The Story of Monsanto. New York: Simon & Schuster, 1977. (285 pp.)
A popular account without citations. The company was founded in 1901.

509. MONTECATINI. The Montecatini Group: History. Products. Development. Milan, 1957. (73 pp.)
Includes two pages of history without references, and is less informative than Montecatini (London: Temple Press, n.d. [1965 ?]) an evaluation of the company by the "Temple Press Economic Intelligence Unit."

510. NOBEL. The History of Nobel's Explosive Co., Ltd. 1871-1926. London, 1938.
Vol. 1 of Imperial Chemical Industries Ltd. and Its Founding Companies, of which I've found no other volumes.

511. NORSK HYDRO. Olsen, Kr. Anker. <u>Norsk Hydro. Gjennom 50 ar</u>. Oslo, 1955. (623 pp.)
The company was founded in 1905, as Norsk Hydro-Elektrisk Kraelstofaktieselskab, to exploit the Birkland-Eyde process for nitrogen fixation. This is a "coffee table book" in style and weight, containing much information and no references. A book with the same title was published in English (Oslo, 1950) with 81 pages mostly devoted to illustrations.

512. PACIFIC COAST BORAX CO. <u>The Story of the Pacific Coast Borax Company</u>. Los Angeles, 1951. (58 pp.)
"The story of five companies" of this name, since 1899, published by their successor. No citations but a narrative remarkable, by the standard of company histories, for its awareness of what history is.

513. PENNSYLVANIA SALT CO. Leavitt, Robert K. <u>Prologue to Tomorrow: A History of the First 100 Years of the Pennsylvania Salt Mfg. Co</u>. Philadelphia, 1950. (100 pp.)
Although some information on the history of company is unavoidable, this reads like an after-dinner speech to the sales force. No citations.

514. PILKINGTON. Barker, T.C. <u>The Glassmakers</u>. London: Wiedenfield & Nicholson, 1977. (557 pp.)
A serious history of the business. Citations. This is the second edition of his <u>Pilkington Bros. and the Glass Industry</u> (London, 1960), which carried the story to 1918 (and had only 296 pp.). The firm originated in 1826, out of an investment in the St. Helens Crown Glass Co.

515. REYMERSHOLMSBOLAGET. Althin, Torstin. <u>Reymersholmsbolaget. Historik</u>. Hälsingborg, 1955. (161 pp.)
History of a Swedish producer of minerals and heavy chemicals. No references.

516. ROHM & HAAS. <u>Chemicals for Industry</u>. n.p. (Philadelphia ?), 1955. (188 pp.)
Privately printed. Principally useful for fifteen pages on the history of the firm. No citations.

517. SAINT-GOBAIN. Choffel, J. <u>Saint-Gobain: du miroir à</u>

l'atom. Paris, 1960. (145 pp.)
Founded in 1665 to manufacture glass, the company added artificial soda in 1806 and subsequently other chemicals. Well-written, from company records but without citations.

518. SAINT-GOBAIN. Compagnie de Saint-Gobain. 1665-1965. Paris, 1965. (114 pp.)
A picture book in which history is represented with splendid reproductions of old illustrations.

519. SAINT-GOBAIN. Pris, Claude. Une grande enterprise François sous l'ancien-régime, la manufacture royale des glaces de Saint-Gobain. Lille, 1975. (2 vols)
Comprehensive, without citations but with an extensive bibliography indicating many unpublished sources.

520. SCHOTT. Kuenert, Herbert, ed. Briefe und Dokumente zur Geschichte der VEB Optik Jenaer Glasswerk Schott. Jena, 1953. (2 vols)
The letters and documents in question date from 1882 to 1886 and are concerned with the establishement of this famous firm.

521. SHELL (Netherlands). Forbes, Robert J., and D.R. O'Beirne. The Technical Development of the Royal Dutch Shell, 1890-1940. Leyden, 1957. (670 pp.)
An exemplary history with citations and attention to petrochemicals by an employee of the company who was a well-known historian of technology.

522. SHELL (United States). Beaton, Kendall. Enterprise in Oil: A History of Shell in the United States. New York, 1957. (815 pp.)
A serious history with citations of a company dating from 1912. Substantial attention to chemicals.

523. SKANSKA CEMENT AKTIEBOLAGET, 1871-1931. Minnesskrift. Uppsala, 1932. (219 pp.)
On Portland cement manufacture in Sweden from company archives. No citations.

524. SOLVAY. Bolle, Jacques. Solvay, l'invention, l'homme, l'enterprise industrielle, 1863-1963. Brussels, 1963. (176 pp.)

Informative although journalistic and without indication of sources.

525. TENNANT. A Short Account of the Tennant Companies, 1797-1922. London, 1922.
By 1835 St. Rollox was the most important chemical works in the world. The Tennant companies reorganized in 1885 and became part of United Alkali company in 1890. This is mainly a pious account of Boards of Directors.

526. TENNANT. Tennant, E.W.D. "The Early History of the St. Rollox Chemical Works," Chemistry and Industry, (1947), 667-73.
Founded in 1794 to exploit bleaching inventions of Charles Tennant. The history is here succinctly recounted by a descendent, with a good eyewitness description of its condition in 1861.

527. TSCHUDI. Tschudi, P. Hundert Jahre Türkischrotfärberei, 1829-1928. Geschichte der Firma Johann Caspar Tschudi. Glarus, 1931. (89 pp.)
Not seen.

528. UNILEVER. Wilson, Chas. H. The History of Unilever. London, 1954. (3 vols)
An outstanding example of the history of a firm. Citations.

529. UNITED ALKALI CO. Centenary of the Alkali Industry, 1823-1923. One Hundred Years of Scientific and Industrial Progress. Widness, n.d.
A mixture of anecdotal material from family archives (some unrelated to alkali) with a narrative "record of the triumphant overcoming of difficulties, of the romance of waste recovery, and of the ultimate suppression of the original process in the inexorable advance of science." No citations.

530. UNITED STATES RUBBER CO. Babcock, Glenn D. History of the United States Rubber Company. A Case History in Corporate Management. Bloomington, Ind., 1966.
(477 pp. Indiana University Institute of Business, Rept. no. 39)
Contains considerable material relating to the technologies involved. Citations.

531. VORSTER & GRUENEBERG. Festschrift der Feier des 50 Jährigen Bestehens der Firma Vorster & Grüneberg. n.p., 1908.
Not seen. This firm, one of the early Kali manufacturers, was known from 1892 as Chemische Fabrik Kali.

532. WASAG (Westfälisch-Anhaltischen Sprengstoff AG). Fischer, Wolfram. WASAG, die Geschichte eines Unternehmens, 1891-1966. Berlin, 1966 (253 pp. Schriften zur Wirtschafts und Sozialgeschichte, Bd. 4)
WASAG began as a gunpowder and explosives maker, expanded into fertilizer and plastics. Citations.

533. WETHERILL. Hussey, Miriam. From Merchants to "Color Men," Five Generations of Samuel Wethermill's White Lead Business. Philadelphia, 1956. (149 pp. Wharton School Research Studies XXXIX)
The technology is minimal as indeed it actually is in the industry. Citations.

CHEMICAL HAZARDS

EARLY WORKS

534. EVELYN, John. Fumii fugium, or, the Inconvenience of the Aer and Smoak of London Dissipated. London, 1661. (33 pp., reprinted in Elmsford, N.Y.:Maxwell, 1969).
He recommends planting beds of "fragrant flowers," adding that the College of Physicians "esteem coal smoke rather preservative against infections."

535. RAMAZZINI, Bernardino. De morbis artificium. Modena, 1700. Eng. trans. as Treatise on the Diseases of Tradesmen, London, 1705 (reprinted by the New York Academy of Medicine, 1964. 483 pp.).
Discusses 51 occupational diseases, those involving chemical hazards including "chemists," gilders, potters, glass, lime and sulphur workers, tanners, apothecaries, vintners and brewers, starch-makers, salt-makers, confectioners and soap-makers. A classic work.

536. HOFFMAN, Friedrich. De metallurgiam morbiferam.

Halle, 1705. (36 pp.)
A brief description with citations of the maladies found in the mineral industries.

537. HENCKEL, Johann F. <u>Medizinische Uffstand</u>. Leipzig, 1745.
On diseases of miners (Bk. I Bergsucht) and metallurgists (Bk. II Hüttenkatze). No citations nor much reference to specific substances but clearly aims at the whole of the mineral industries.

538. (TILLET, [Mathieu?] and - FORGEROUX). "Sommaire des observations faites par ordre du Roi, sur les cotes de Normandie, au sujet des effets pernicieux qui sont attributés dans le pays de Caux, à la fumée de varech, lorsqu'on brule cette plante pour la réduire en soude," <u>Observations sur la Physique</u>, 2 (1772), 313-23.
A response to a claim that the burning of varech (kelp, for the production of alkali) caused epidemic illness, killed grain and fruit, deprived the farmers of fertilizer and fish of spawning ground. The article claims that the smoke doesn't bother anyone. It denies all except the value of varech as fertilizer. They report that (J.E.) Guettard had made a similar study in the Mediterranean with the same conclusion. The king had authorized the varech burning in 1731.

539. HAYES, Thomas. "On the Danger of Using Vessels of Lead, Copper, or Brass in Dairies," <u>Repertory of Arts and Manufactures</u>, 7 (1797), 116-21.
"Conjectures" that claims for "badness" in dairy products may stem from the vessels used.

540. GUYTON de MORVEAU, L.B., and J.A. CHAPTAL, "Rapport... sur la question de savoir si les manufactures qui exhalent une odeur désagréable peuvent etre nuisable à la santé," <u>AC</u>, 54 (1805), 86-103.
Response to a query from the government to the National Institute inspired by "arbitrary" legislation which had banished many factories from cities. Divides them into two classes, (1) those producing emanations, from putrefaction or fermentation, deemed nuisances or dangerous, and (2) emanations from processes involving heating, which are "more or less"

disagreeable and injurious. To the first class belong works for flax processing, slaughter houses, tanneries, starch, and breweries; to the second distillers of acids, spirits, animal substance, works for gilding, lead, copper and mercury preparations. They absolve all but hemp processers and establishments where "a large quantity of animal or vegetable matter is subjected to humid putrefaction." But all are to be watched and some licensed. It is evident that the concern is to protect manufacturers who have been under general attack.

541. PELLITAN, P. "Essai sur les moyens de retenir l'acide muriatique qui se dégage pendant la décomposition en grand du sel marin par l'acide surfurique." AC, 75 (1810), 176-93.
Says he worked with Holker at Rouen and recognized from the outset that hydrochloric acid could not be vented in the neighborhood. Describes attempts to absorb it in water. Says they absorb it by passing the fumes alternately through two "canals" filled with limestone. Describes the apparatus.

542. DARCET, J.P.J. "Description d'un appareil au moyen duquel on peut éviter toute mauvaise odeur, dans la fabrication du bleu de Prusse," AC, 82 (1812), 165-70.
Describes a vessel in which the openings through which gases might escape are sealed.

543. FODERE, F.E. Traité de médicine légale. Paris, 1813. (6 vols)
Polution problems are dealt with in 6:317-30, with numerous examples. He particularly deplores mineral acid factories. A sulphuric acid works at Marseilles exhaled the acid "from all its parts" and has desolated the fields around (but it was forced to close in 1810). Thirty artificial soda works in the same place fill the atmosphere with "all kinds of gases." On the other hand, a method has been devised to free coal from sulphur. He condemns ignorant workmen and unresponsive courts.

544. GUYTON de MORVEAU, L.B., and J.A. CHAPTAL, "Extrait d'un rapport fait a l'Institut de France...sur la

question de savoir si les manufactures qui exhalent une odeur désagréable peuvent etre nuisable à la santé," BSE, 13 (1814), 42-7.

Sequel to an earlier report (no. 540), concluding, in general, that organic industries are noxious, inorganic industries not--with some exceptions. The following year it was reported (Ibid. 14:66-71) that it was proposed to classify industries into those which must be located away from habitations, those where it isn't "rigorously necessary," and those which pose no inconvenience. A decree was published in which Prussian blue manufacture was put in the first class, but sal amoniac and artificial soda into the third.

545. PAJOT DESCHARMES, C. "Mémoir sur les moyens de remédier aux inconvéniens occasionés par les vapeurs ou gaz délétètres qui s'élèvent des fabriques de soude artificielle. Annales de l'Industrie Nationale et Etrangeres, 21 (1826), 267-97.

Response to a prize offered by the Société des Amis des Sciences at Aix, containing various suggestions, including absorption and tall chimneys. It received honorable mention.

546. FRANCE. "Rapport générale sur les traveux de conseil de salubrité pendant l'année 1825," BSE, 7 (1827), 382.

The council discussed 230 reports, including 22 on metal works, 16 on dye works, 19 on steam engines, 11 on "chemical works."

547. THACKRAH, C.T. The Effects of Arts, Trades, and Professions, and of Civic States and Habits of Living, on Health and Longevity. London, 2nd ed., 1832. (238 pp.)

Chemical problems are lumped with others. His "classes" of industry are "not decidedly hurtful," "apparently beneficial," and decidedly injurious." Vegetable exhalations are in the first; animal exhalations (notwithstanding the odors) are in the second. The third includes "peculiar atmospheric impurities" from lead, zinc, sal ammoniac, sulphur, gas works (sulphuretted hydrogen), coke fumes and carbonic acid gas.

548. POPPE, J.H.M. Die Kunst, Leben und Gesundheit der Handwerker, Künstler, Fabrikanten und anderer Handarbeiter soviel wie moglich vor den Gefahrens ihres Lebens su sichern. Heilbron, 1833. (112 pp.)
Intelligently divided into hazards, atmospheric, physical, dangers of falling, explosions, fire, etc. All the chemical trades fall under one or the other. Useful as a guide to what chemical trades existed although he fails to mention artificial soda manufacture.

549. FRICK, -. "Gutachten über die Frage: ob die von Hernn Philipp Henrich Pastor...Sicherheitsvorrichtung..." Verhandlungen des Vereins zur Beforderung des Gewerbfleisses in Preussen (Berlin), 13 (1834), 51-4.
Deals with the question of whether Pastor's safety apparatus against steel and rock dust is also good for protection against white-lead dust. Inconclusive but at least he cares.

550. LOMBARD, Henri C. De l'influence des professions sur la durée de la vie. Geneva, 1835.
Not seen.

551. DE LA BECHE, H.T., and Lyon PLAYFAIR. Upon the Means of Obviating the Evils Arising from the Smoke Occasioned by Factories and Other Works Situated in Large Cities. London, 1846. (Great Britian, House of Commons, Sessional Papers, vol. 43, no. 194)
Not very conclusive. They emphasize the difficulties, concluding that house fires are the major evil and the investigation of "other works" impractical. Some towns already have regulations and factory inspectors.

552. SPENCE, Peter. Coal, Smoke and Sewage, Scientifically and Practically Considered. Manchester, 1857. (32 pp.)
He has a solution to the problems of disease-carrying sewage, black smoke, and invisible effluvia, namely "a system of gaseous sewage" in which the effluvia are led into the town drainage, where the "sulphureous acid gas" would neutralize the sewage into "ammonium sulphite," and the gases would finally be led to a 600 ft. chimney. The latter would require 18.2 million bricks. The scheme was replete with such calcula-

tions, in which he was assisted by Angus Smith, subsequently the high-priest of the British Alkali Acts.

553. SMITH, R. Angus. "On the Air of Towns," <u>Journal, Chemical Society (London)</u>, 11 (1859), 196-235.
Aims to rectify the circumstances that "neither those who made the laws (against the atmospheric pollution) nor those who administer them have ever taken the pains to find out what it really was against which they combatted." He finds Manchester ideal for such a study, and gives analyses for sulphur, acids, ozone, carbon dioxide, organic matter, ammonia, carbon, and "tarry matters."

554. GREAT BRITAIN. <u>Report of the Royal Commission on Noxious Vapors.</u> London, 1878. (589 pp. Parliamentary Papers C 2145, vol. 44)
A 37-page report is followed by minutes of evidence. The committee visited the sites of manufacture of alkali, cement, "chemical manure," coke ovens, works for copper, lead, nickle, glass, salt and pottery. The recommendations relate mainly to the conduct of the Alkali Acts.

555. GREAT BRITAIN, Local Government Board, <u>Alkali Act. Annual Report of the Inspector.</u> London, 1865 ff.
This report, of which no. 104 appeared in 1974, is something of a running account of the state of the British heavy chemical industry. The first Inspector was Robert Angus Smith, and his first report (for 1864) explores the industry back to 1837, thus nearly to the beginning of the artificial soda industry.

HISTORIES

556. <u>A Full Report of the Trial of the Important Indictment Preferred by the Corporation of Liverpool against James Muspratt, Manufacture of Alkali.</u> London, n.d. (1838 ?)
The indictment was "for creating and maintaining a nuisance," and the trial is here chronicled by a supporter of Muspratt, who says that the reader will find "the chemical talents of Dr. (Thomas ?) Thomson" superior to the Corporation's "chemical witnesses." The judge noted, however, that the majority of Muspratt's witnesses lived around the factory and gained

their livelihood from it and added "when scientific evidence is got up for any purpose, it must always beget a bias." Muspratt was judged guilty. In addition to the trial this is something of a history of the early British alkali industry.

557. "The History and Working of the Alkali Acts," Chemical News, 38 (1878), 157-60, 181-82, 201-02.
Summarizes a report of the Noxious Vapors Commission (no. 554, above). The author (probably William Crookes, Editor of Chemical News) reports that the acts, passed in 1862 and overseen by R. Angus Smith, were working well. The second part describes the causes which led to the acts, and the third describes current problems.

558. AYNSLEY, E.E., and W.A. CAMPBELL, "John Glover and the Clean Air Acts," Chemistry and Industry, (1959) 1540-41.
A capsule account of Glover's life and of the problem he solved. The first "Glover tower" was erected in County Durham in 1859.

559. BIASUTTI, G.S. Histoire des accidents dans l'industrie des explosifs. Montreaux: Corday, 1978.
Not seen.

560. COKER, W.S. "Spanish Regulation of the Natchez Indigo Industry, 1794-94: The South's First Anti-Pollution Laws?" T&C, 13 (1972), 55-58.
Indigo production was prohibited because of its pollution of streams on which cattle depended; but the effectiveness of the regulation is unclear because of other adverse effects on the industry, stemming from bad weather and the growth of a rival industry, cotton. Citations.

561. FLICK, Carlos. "The Movement for Smoke Abatement in 19th Century Britain," T&C, 21 (1980), 29-50.
On the continuation of a movement spanning "almost seven centuries." Describes devices and concludes that decisions were left to politicians and that "little abatement was achieved."

562. GOLDWATER, L.J. "From Hippocrates to Ramazzini: Early

History of Industrial Medicine," <u>Annals of the History of Medicine</u>, 8 (1936) 23-35.
Cites Hippocrates on the diseases of fullers (who cleaned clothes by working them in urine), Martial's description of the diseases of sulphur workers, and Galen's of those of workers in gypsum and copper, the latter observed on his visit to the mine on Cyprus.

563. GRINDER, R.D. "<u>The Anti Smoke Crusades: Early Attempts to Reform the Urban Environment, 1893-1918.</u> Dissertation, Univ. of Missouri, 1974. (Univ. Microfilm 74-18539)
Not seen.

564. HAMILTON, Alice. <u>Exploring the Dangerous Trades</u>. Boston, 1943. (433 pp.)
This autobiography of a physician who began the study of "occupational diseases" in 1910 contains a first-hand description of many chemical industries in the United States at that time. See also her "Industrial Poisoning in American Aniline Manufacturing," <u>Monthly Labor Review</u>, 8 (1919), 517-33.

565. MACAREL, L.A. <u>Manuel des ateliers insalubres et incommodes</u>. Paris, 1828. (306 pp.)
Not seen.

566. MACLEOD, R.M. "The Alkali Acts Administration, 1863-84: The Emergence of the Civil Scientist," <u>Victorian Studies</u>, 9 (1965), 85-112.
Why was centralized control in the Alkali Administration relatively easily accomplished? Attributes it to the position of the subject in a "safety zone" between "public ignorance and national apathy," to a simplified redress and successful encouragement of collective responsibility among the manufacturers, and to the scientific standards and "helpful solicitude" of the inspectors. Citations.

567. MONTEIL, J. "Un procès de pollution industrielle à Montpellier en 1791," <u>Histoire des Sciences Medicales</u>, 8 (1974), 825-27.
On Chaptal's chemical works.

568. NORRIS, W.G. <u>Chemical Service in Defence of the Realm:</u>

100 Years of Chemical Inspection. London: Ministry of Supply, 1957. (91 pp.)
A narrative account without references of the history of the Chemical Inspectorate, which was established in 1854 at Woolwich Arsenal.

569. Oliver, Thomas, ed. Dangerous Trades: The Historical, Social and Legal Aspects of Industrial Occupations as Affecting Health, by a number of experts. London, 1902. (891 pp.)
After a brief historical introduction the experts deal with lead, arsenic, phosphorus, mercury, bichromates, rubber, dinitrobenzene, benzene, explosives and "the chemical trades." There are few citations and the authors are little interested in history. This is more useful for the state of the art.

570. RODERICK, G.W., and M.D. STEPHANS. "Profits and Pollution," Industrial Archeology, 11 (1974), 35-45.
On the Muspratt suit (see item 556).

571. SMITH, Robert Angus. River Pollution Prevention Act of 1876. London, 1882.
This has been described as his reminiscences and so should be important. But the only library where I have found it listed (New York Public Library) can't find it.

572. TE BRAKE, W.H. "Air Pollution and Fuel Crises in Pre-industrial London, 1250-1650," T&C, 16 (1975), 337-59.
Deforestation and the introduction of coal as fuel created an environmental crisis which "was resolved by a population reduction of 40% during the 14th century."

573. WICKERSHEIMER, E. "Fumées industrielle et établissements insalubres à Rouen, 1510," Annales de Hygene Publiques, 5 (1927), 567-75.
Describes an inquiry of Sept. 24, 1510, inspired by the introduction of coal for industrial firing.

PART II

ACIDS AND ALKALIS

EARLY WORKS

Vinegar (acetic acid) was known in antiquity, the mineral acids (nitric, sulphuric, and hydrochloric), in varying degrees, from the late Middle Ages. Production methods were fairly standardized, the principal novelty being the production of sulphuric acid from the fumes of burning sulphur, which appeared in the seventeenth century. All this was included in any chemical text of any generality, and there was little special literature on the acids. A few early works are mentioned below, none of which equal the 58 pp. on this topic in Demachy's <u>Distillateur...d'eaux-fortes</u> (1776, no. 18) in the information supplied. By contrast, the alkalis were less well identified, were thought to be more various, and hence gave rise to a larger special literature.

574. BORELLI (Pierre Borel ?). "Manière proposée pour tirer beaucoup d'esprit de souffre," <u>AdS Paris</u>, 1 (1733), 372.
A half-page note dated 1683, proposing an improvement whereby a half-filled retort of water is to have a side-tube, to admit the "fume" of sulphur. (NOTE. The first several volumes of this periodical were published simultaneously and print earlier work. After these there were no volume numbers.)

Figure 2. Apparatus for distillation of iron vitriol to produce "Nordhausen" (fuming) sulphuric acid. From Figuier (item 112)

575. HOMBERG, Wilhelm, "Comment tirer beaucoup d'esprit acid de souffre," AdS Paris, 2 (1733), 109. (dated 1694)
Reads (in its entirety), "One draws more spirit of the acid of sulphur in a cave if one makes a hole in it; and still more if the hole is made in a mass of snow."--which at least suggests more than laboratory production.

576. JUSSIEU, Antoine de, "Histoire du kali Alicante," Ads Paris, (1717), Mém., 73-78.
Primarily a botanical study but reveals the confusion of names. He says that the French use "kali" for the alkali from many plants, but "soude" for this one and that it is identical to "barilla."

577. DUFAY, (C.F. ?), "Sur la fabrique de la potasse," Ib., (1727) Hist., 34-36.
A description of potash production, which he says is hardly known, in the forests of the Moselle. The calcined product is called salin.

578. DUHAMEL DU MONCEAU, H.L. "Sur la base du sel marin," Ib., (1736), Mém., 215-32.
The pioneering work in differentiating soda from potash.

579. MITCHELL, John. "An account of the preparation and uses of the various kinds of potash," PT, 45 (1748) 541-63.
Refers especially to America. He decries the general misunderstanding (which he shares) of the nature of the substance.

580. WARREN, Peter. "How to Make the Best Russia Potashes Gentlemen's Magazine, 23 (1753), 400.
A circumstantial description. Russia was a principal supplier of potash to western Europe at that time.

581. STEPHENS, Thomas. The Method and Plain Process for Making Potashes Equal, If Not Superior to the Best Foreign Potash. London, n.d. (1775 ?). (36 pp.)
Inspired by acts of Parliament (1751) to encourage potash production. Wood is the source. His method is not remarkable. Stephens reportedly had a "potasherie" in South Carolina in 1747.

582. STEPHENS, Thomas. <u>The Rise and Fall of Potash in America</u>. London, 1758. (43 pp.)
Not seen.

583. DOSSIE, Robert. <u>Observations on the Pot-ash Brought from America</u>. London, 1767. (41 pp.)
A sample was received in connection with an application from the House of Representatives of the Province of Massachusetts for a premium offered by the Royal Society of Arts. He evaluates it and suggests improvements.

584. LEWIS, Wm. <u>Experiments and Observations on American Potash</u>. London, 1767. (34 pp.)
Another book of comments on the Massachusetts potash; this one published by the Society of Arts. He judges it less favorably than did Dossie, also suggests improvements and describes a good assay.

585. GLEDITSCH, J.G. "Relation succinct concernant la terre de Debrezin," <u>AdS Berlin</u>, Nouveaux Mém. (1770), 8-18.
Dry-lake deposits of a "mineral alkali" resembling Egyptian "natron" were exploited near this Hungarian town at the end of the eighteenth century.

586. MONRO, Donald. "On a Pure Native Crystallized Natron, or Fossil Alkaline Salt, Which Is Found in the Country of Tripoli in Barbaby," <u>PT</u>, 61 (1771), 567-73.
Describes a sample obtained from the widow of the Consul in Tripoli, who reported that large quantities were brought annually from the interior, under the name <u>trona</u>.

587. ROSEN von ROSENSTEIN, Nils. <u>Diss. Inaug. Chemica de Genesi et Ortu Salis Alkali Fixi Vegetabilis</u>. Argentorati, 1776. (128 pp.)
Not seen. According to a reviewer he gives a complete history of opinions on vegetable alkali. There was also a German edition.

588. CAMPBELL, Alexander. <u>Diss. Med. Inaug...de acido vitriolico</u>. Edinburgh, 1778. (30 pp.)
Not seen.

589. "Ueber die Bereitung der verschiedenen, im Handel gangbaren, Arten Scheidewasser in den Fabriken," Handlungszeitung, 1 (1784), 329-30.
Description of nitric acid production.

590. WILDENHAYN. Abhandlung vom Pottaschsieden. Dresden, Th. 1, 1786, Th. 2, 1800. (90 pp.)
A comprehensive account of production in central Europe. The author, who had his own factory, was winner of a potash prize offered by the Leipzig Economic Society in 1764.

591. "Kurzen Geschichte der drey mineralischen Saüren, nebst ihren vornehmsten Produkten," Handlungszeitung, 4 (1787), 34-36, 44-46, 52-55.
Describes conventional methods of production of the three mineral acids.

592. WIEGLEB, J.C. "Ueber das wahre Verhaltniss der Saüre im Schwefel," (Crell's) Chemische Annalen, 2 (1791) 400-13.
Includes a history of research, which he says has been going on for a century.

593. MEIDINGER, Karl von. "Ueber den Goldscheidungsprocess zu Nagybancen und Kremnitz in Ungarn," Journal für Fabrik, 14 (1798), 1-22.
Describes the varieties of nitric acid used in separating gold and silver and gives cost data on production.

594. ANDREOSSY, A.F. "Particulars Concerning the Valley of the Natron Lakes," and BERTHOLLET, C.L., "Observations on the Natron," in Memoirs Relative to Egypt. London, 1800, pp. 253-69.
Descriptions of the Egyptian dry-lakes which had been sources since antiquity of a very impure mineral soda. Trans. from France. Commission des Sciences et Arts d'Egypt, Mémoires sur l'Egypt pendant les campaigns... Paris, 1800 ff.

595. "Programme d'un prix de technologie proposée par l'academie royal des sciences de St. Petersbourg... 29, Dec., 1829," Annales de l'industrie Française et Etrangère, 5 (1830), 339-47.

Describes Russian sources of soda and potash, the shortage of the former and the decimation of forests to produce the latter. Recommends adoption of the Leblanc process.

596. ANDERSON, J.D. The Barilla Question Discussed.
     Edinburgh, 1831.
     On the manipulation of duties to save the Scottish kelp-burning alkali industry from the English "barilla" industry, the latter referring to "the black ash gentry" who make Leblanc soda. Good analysis of the advantages and disadvantages of various locations for soda works. The author, who says that some people make Leblanc soda as a cottage industry, himself manufactured Leblanc soda near Leith.

HISTORIES

597. BODMAN, Gösta. "The Production of Potash in Sweden until the Middle of the 18th Century," Daedalus (Stockholm), 1950, 89-120. (in Swedish)
     Straightforward historical account from Peder Månsson (d. 1534) to the end of the traditional industry.

598. BROWNE, Chas. A. "Historical Notes upon the Domestic Potash Industry in Early Colonial and Later Times," JCE, 3 (1926), 749-56.
     Deals with North America up to 1811, very briefly, but incorporating evidence not found elsewhere. Citations.

599. CLOW, A., and N. "The Natural and Economic History of Kelp," AS, 5 (1947), 297-316.
     Spanish barilla (soda from seashore plants) was the most popular alkali in Britain, which exempted it from embargo even when at war with Spain in 1717. But "seaweed ash" (soda from kelp) began to compete with it in the eighteenth century.

600. GILLENS, L. "Premiums for Vegetable Alkali...1758-1827," Journal Society of Arts (London), 103 (1963), 577-81.
     Describes the prizes which inspired a considerable literature and some potash production in North America.

601. GLASSORD, C.F.O. "History and Description of Kelp Manufacture," Proceedings, Philosophical Society of Glasgow, 2 (1844-45), 241-60.
Produced in Scotland for about a century as a source of alkali. Describes economic vicissitudes and a revival after 1845 because of a demand for iodine.

602. KLINCKENSTROEM, Carl G. von. "Soda, Versuch einer Bibliographie," Börsenblatt für den Deutschen Buchhandel, 19 (1963), 113-15.
Nineteen references in early technological treatises to descriptions of the production of natural soda.

603. KREPS, T.J. "Vicissitudes of the American Potash Industry," Journal of Economic & Business History, 3 (1931), 630-66.
Brings the story of potash in the United States into the twentieth century with particular attention of data on production and trade.

604. LECLAIR, E. La fabrication des acides forts à Lille avant 1790. Poitiers, 1901.
Not seen.

605. MAYER, Ernst. "History and Modern Aspects of Vinegar Making," Food Technology, 17 (1963) May, 75-78.
Very summary. No references.

606. MULTHAUF, R.P. "Potash in the United States," in Brooke Hindle, ed., Material Culture of the Wooden Age, Tarrytown, N.Y.: Sleepy Hollow Press, 1981, pp. 227-40.
A summary history. Citations.

607. ROBERTS, W.I. "American Potash Manufacture before the American Revolution," Proc. APS, 116 (1972), 383-95.
An account of attempts by English promoters, largely unsuccessful, to establish a potash industry. Appends a description of Thomas Stephens' process, from Gentlemen's Magazine, 1755.

608. SPETER, Max. "Zur Geschichte der Pottasch und ihres Namens," AGNT, 2 (1910), 201-13.
Disproves the idea that it comes from the chemist, J.H. Pott. Citations.

609. SZOKEFALVI-NAGI, Z. "The Technology of Hungarian Potash Production in the 18th Century," Technika-förtenet Szemle (Budapest), 2 (1964), 110-16.
(in Hungarian, with summaries in western languages.

ALUM (Aluminum sulphate of potassium or ammonium) and VITRIOL (sulphates of iron)

Conspicuous for their large crystals and astringent taste, these substances were known in prehistoric times, and their mildly corrosive character made them useful. The alums were particularly used for fixing dyes to cloth (mordanting), ferrous sulphate for producing a black color when mixed with a solution of gall nuts.

EARLY WORKS

610. MINDERER, Raymond. De calcantho seu vitriolo. Vienna, 1617. (113 pp.)
Appears to be the first book on this chemical genre.

611. CANEPARI, P.M. De atramentis cijuscunque generis. Venice, 1619. (568 pp.)
Much dependent on Georg Agricola but primarily an attempt to unravel the nomenclature for the benefit of medicine. The name atrament (blacking) was often used for ferrous sulphate, which was used in making black pigment.

612. MORAY, Sir Robt. "On the mineral of Liège, yielding both brimstone and vitriol, and the way of extracting them out of it used at Liège," PT, 1 (1665), 45-6.
A circumstantial description, presumably exemplifying the Royal Society's effort to describe "trades."

613. TALBOT, G. "A Description of a Swedish stone which affords sulphur, vitriol, alum, and minium," PT, 2 (1666), 375-6.

614. COLWALL, Daniel. "An Account of the English Alum Works," PT, 12 (1679), 1042-59.
Describes alum and vitriol manufacture in Yorkshire and Lancashire.

615. GEOFFREY, E.G. "Détail de la manière dont on fait l'alun de Roche en Italie & en Angleterre," AdS Paris, 1702, Hist. 20-22.

616. LEMERY, Louis. "Nouvel éclaircissement sur l'alun, sur les vitriols, et particulièrement sur la composition naturelle & jusqu'à present ignorée," AdS Paris, 1735, Mém., 262-80.
The most important early attempt to elucidate the chemical character of the alums and vitriols. But not the first: in introducing this, Fontanelle, Perpetual Secretary of the Academy remarks, "what material has been more examined and more tormented by the chemists than vitriol?" Lemery published a second memoire in the same place (385-402) and a third in 1736 (Ib., mém. 265-301).

617. BERGMAN, T.O. De confectione aluminis. Uppsala, n.d. (1767 ?). (English trans. in his Physical and Chemical Essays, London, 1788, vol. 1, 338-96).
On the history of alum, its sources, production, uses, and "principles." An essay by a master (although he reports in AdS Stockholm, 38 [1776] 179-90) that he has found the De confectione to be partly in error).

618. MAZEAS, Abbé G. "Mémoire sur les solfatares des environs de Rome: sur l'origin & la formation du vitriol Romain," AdS Paris, Mém. de Mathématique & de Physique, 5 (1768), 319-30.
Circumstantial account of a visit to Tolfa in 1758.

619. MONNET, A.G. Traité de la vitriolisation et de l'alunation, ou l'art de fabriquer l'alun et le vitriol. Amsterdam, 1769.
Treats the manufacture of vitriol and alum in some detail with particular reference to Liège.

620. ENGSTROM, E. "Vom Alaunmachen," AdS Stockholm, 36 (1774), 279-300.
An account by a Swedish alum-maker of his experiments to improve its crystallization.

621. ERXLEBEN, J.C.P. "Chymische Untersuchung der Gravenhorstischen rothen Alauns," Berliner Sammlung zur Beforderung der Arzneiwissenschaften, 8 (1776), 63-70.

Also in his Physikalisch-chemische Abhandlungen, Göttingen, 1776, vol. 1; 304-29.
Discusses the product of the Gravenhorst brothers at Brannschweig, one of the earliest well-known German chemical manufactories (chiefly because of their innovations in advertising).

622. HAGEN, T.V. van der. Beschreibung der Stadt Freyenwalde, das dasigen Gesundbrunnens und Alaun-Werks. Berlin, 1784. (124 pp.)
One of the earliest detailed descriptions with illustrations of an alum works is found on pp. 91-112. It commenced operations in 1718.

623. WILLEKOP, J.J. "Einige Bemerkungen über das Allendorfische Salzwerk, den Weissner und die an demselben gelegenen steinkohlen Bergwerke; und über die Tiegelfabriken und Alaunwerke zu Gross Allmerode," (Crell's) Beiträge zu Chemische Annalen, 2 (1787), 476-94.
Includes a detailed description of a vitriol works which was part of a small complex of mines and chemical factories.

624. KOEHLER, A.W. "Bemerkungen des Vitriol- und Schwefelwerk zu Schreiberhau in Niederschlesien," Neues Bergmannisches Journal (Freyberg), 1 (1795), 564-76.
Interesting description of a factory which made vitriol, sulphur, and sulphuric and nitric acids.

625. DUPQUET, -. "Mémoire sur les terres sulfurique de Rollat, Dept. Somme, et sur la manufacture de sulfat d'alumine établis dans cette commune," Jour. des Mines, 4 (1796), 49-59.
Circumstantial description of the factory as of the date of publication. The chemist, Vauquelin, comments on this work in the same issue, pp. 74-77.

626. VAUQUELIN, L.N. "Mémoire sur la nature de l'alun du commerce," Jour. des Mines, 5 (1796-97), 429-44.
A masterful analysis, particularly of the difference between ammonium and potassium alums.

627. CAVILLIER, -. "Mémoire sur les aluminières du pays de Nassau-Saarbruck," Jour. des Mines, 8 (1798), 763-88.

Good descriptions of the alum works of the Saar region, active since 1715 and now (temporarily) under French control.

628. "Prix pour la fabrication de l'alun," BSE, 1 (1802), 105-06.
The Société d'Encouragement pour l'Industrie Nationale (Paris) announces a prize for an explanation of why Roman (i.e. Tolfa) alum is considered superior. The sequel, reported in BSE, 3 (1804) 155-59, 214; and 4 (1805) 58, 129, 222, and 231, show that many of the leading French chemists participated, and that F.A. Curaudau won the prize. Whatever the case with Roman alum, it is finally reported that France now imports little alum and that she is also self-sufficient in vitriol, thanks to new processes. The German journal, Magazin für die neuesten Zustand der Naturkunde, 12 (1806), 325-77, gives even more detail on all this.

629. KLAPROTH, D.M.F. "Chemisches Untersuchung des Alaunsteins von Tolfa und des erdigen Alaunschiefens von Freienwalde," Neues allgemeines Journal der Chemie, 6 (1805), 35-54.
Important, with much historical information. Trans. into French and English (Nicholson's Journal of Natural Philosophy, 20 (1808), 359-71).

630. "Mines d'alun et aluminieries existant ou susceptable d'être mises en activité dans le Dept. de L'Aveyron," Jour. des Mines, 19 (1806), 120-34.
Correlates production with the geological-mineralogical conditions of the ore used.

631. (KNOX -). "Some Account of a Very Singular and Important Alum Mine near Glasgow," (Nicholson's) Journal of Natural Philosophy, 16 (1807), 232-34.
Describes one of the most important British works at Hurlett.

HISTORIES

632. BECK, Romayn. "Memoire of Alum," Trans., Society for the Promotion of Agriculture, (New York), 4, 1 (1816), 50-92.
A history, with reference to American sources.

633. CLOW, Archibald, & Nan CLOW. "Vitriol in the Industrial Revolution," Economic History Review, 15 (1945), 44-55.
Mistitled; it is really a history of the production of sulphuric acid ("oil of vitriol") in England up to about 1840. Citations.

634. DELUMEAU, Jean. L'alun de Rome, 15-19 siècle. Paris, 1960.
Not seen.

635. HEERS, J-L. "Les Genois et le commerce de l'alun à la fin du moyen age," Revue Historique, Economique, et Sociale, 32 (1954), 31-53.
Trade and production data from the 15th century. Citations.

636. JENKINS, Rhys. "The Alum Trade in the 15th and 16th Centuries and the Beginning of the Alum Industry in England," in Rhys Jenkins, Collected Papers, Cambridge, 1936: 193-203 (reprinted from Trans. of the South East Union of Scientific Societies).
Dates the beginning of the British industry 1564.

637. LEVY, Martin. "Alum in Ancient Mesopotamian Technology," Isis, 49 (1958), 166-69.
Somewhat speculative.

638. SINGER, Charles. The Earliest Chemical Industry. London, 1948. (337 pp.)
A history of alum, elegantly illustrated and comprehensive, perhaps definitive.

639. SZOEKEFALVI-NAGY, Z. "Alum Production in Hungary in the 18th Century," Technikaförtenet Szemle (Budapest), 4 (1967), 123-34.
First made at Parad in 1763. In Hungarian with summaries in other languages.

640. TURTON, R.B. The Alum Farm. Whitby, 1938. (203 pp.)
Comprehensive description of manufacture at Whitby, 1612-25.

BORAX

Known in Islam from the 10th century or earlier, borax was used in soldering for fine metal work. That such a use made it an important industrial chemical is a measure of the scale of early industrial chemistry. A rare material; it was successively a monopoly of Tibet, Tuscany (from c. 1780) and California (a near monopoly from c. 1860).

EARLY WORKS

641. LEMERY, Nicolas. "Sur le borax," AdS Paris, (1703), Hist., 49-51.
An important early attempt at an analysis, inspired by an earlier essay by Wm. (Guilaume) Homberg in the same journal (1702) and followed by another by Lemery himself (1728, Mém.: 273-88).

642. RENNEWALD, H.C. Diss. inaug. chemico-medico de borace. Halae Magdeburg, 1745. (40 pp.) (dissertation, M. Alberti, praes.)
Good bibliographical introduction to borax.

643. CNOLL, S.B. "Abhandlung von einen indienischen natur-liche alkalinischen Salz, und die Bereitung des Borax in Indien," (Schreber's) Sammlung Verschiedener Schriften, 2 (1756), 288-93.
One of the earliest accounts of "Indian" (actually Tibetan) borax, sent from Tranquebar where Cnoll was a physician to the Austrian mission.

644. HOEFER, F. Memoria sopra il sale seditivo naturale della Toscana. Florence, 1778. (56 pp.) (reprinted in Nasini, no. 664)
Announces the discovery of "natural sedative salt" (boric acid) in lagoons near Sienna; includes a good general discussion and history of borax.

645. WEBER, Jacob A. "Von dem Sedativsalz des Hombergs," Physikalisch-Chymisches Magazin, 2 (1780), 185-242.
Some references to other work but chiefly a description of his own experiments, as profuse as was customary with this chemist.

646. FUCHS, G.F.C. <u>Versuch einer naturlichen Geschichte des Boraxes</u>. Jena, 1784. (96 pp.)
A survey of the literature, noteworthy for its ample annotation.

647. CIBOT, P.M. "Note sur le borax," <u>Memoires concernant l'histoire, les sciences...des Chinois, par les missionaires de Pekin</u>, 11 (1786), 343-46.
Says that the Chinese claim that borax (<u>pong cha</u>) was known "anciently" although he can find no evidence of it in books; continues with other information on its contemporary sources.

648. PROUST, J.L. "Sur le borax," <u>Journal de Physique</u>, 30 (1787), 393-96.
Mentions borax in South America on the basis of a report received from Potosi in 1786.

649. BOWRING, J. "On the Boracic Acid Lagoons of Tuscany," <u>Philosophical Magazine</u>, n.s. 15 (1839), 21-25.
A succinct account of their appearance with production data.

650. PAYEN, Anselm, "On the History and Manufacture of Boracic Acid in Tuscany," <u>Chemical Gazette</u>, 1 (1842), 214-19.
Deals particularly with the technology involved.

651. CUNNINGHAM, Alexander. <u>Ladak, Physical, Statistical and Historical</u>. London, 1854.
Gives an account (pp. 144-45) of a personal visit to the sites of borax production in Tibet.

652. VEATCH, John A. "Notes of a Visit to the Mud Volcanos in the Colorado Desert...July, 1857," <u>AdS California</u> (San Francisco) 1 (1854-57), 120.
Early reference to borax in California.

653. BOLLAERT, Wm. "The Nitrate of Soda and Borate Districts of Peru," <u>Technologist</u>, 1 (1861), 115-20.
A circumstantial account of production, with some history.

654. PHILLIPS, John A. <u>Report on the Property of the California Borax Co.</u> San Francisco, 1866. (12-14 pp.)

Describes the first attempt to produce borax in California at Clear Lake, Lake County. Further detail is given in the Mining and Scientific Press (San Francisco), 12 (1866), 335 and in subsequent issues.

655. CAMPBELL, M.R. Reconnaissance of the Borax Deposits of Death Valley and Mohave Desert. Washington: GPO, 1902.
On the status of various borax deposits in the western United States.

HISTORIES

656. DYER, B.W. "Searles Lake Development," Colorado School of Mines Quarterly, 45 (1950), 39-44.
A good history of the lake, "discovered" by John Searles in 1863 and exploited for borax from 1874 (after Searles had seen Francis M. [Borax] Smith working Teale's Marsh in nearby Nevada for the material). Describes the present situation of the lake, "the largest exposed salt body in the United States."

657. FIUMI, E. L'Utilizzazioni dei lagone boraciferi della Toscana nell'industria medievale, Pubblicazioni delle R. Universita degli Studi di Firenze, Facolta di Scienze Econ. e Comm. 21 (1943)
Not seen.

658. FRANK, Adolph. "Borsäure gewinnung in Toskana," ZAC, 20 (1907), 258-64.
A summary account of its history and processes. No citations.

659. GERSTLEY, J.M. Borax Years, 1933-1961: Some Recollections. Los Angeles: U.S. Borax Co., 1979. (93 pp.)
Personal reminiscences of the first president of the company which emerged from the romantic era of wildcat exploitation of this mineral in the United States. Undocumented but modest and informative.

660. GOWER, Harry F. Fifty Years in Death Valley: Memoirs of a Borax Man. San Bernadino, Calif.: Death Valley 49ers, 1969. (145 pp.)

Anecdotes of one who began in 1909 a half-century association with the Pacific Coast Borax Co. and its successors.

661. HANKS, Henry G. Report on the Borax Deposits of California and Nevada. Sacramento, 1883. (Reprinted in Harold Weight, Twenty Mule Team Days. Twenty Nine Palms, Calif.: Calico Press, 1965)
Includes a careful account of early history.

662. HART, Ed. "Boric Acid and Borax," V International Kongress für angewandte Chemie (Berlin, 1903), Berichte, 1: 772-73.
In the United States, a summary of recent history.

663. MULTHAUF, R.P. "The Discovery of Borax," 13th International Congress of the History of Science (Moscow, 1971), Proceedings, 7: 103-09.
On European efforts to discover the source of this material at the time it was obtained from the orient.

664. NASINI, R. "I soffioni boraciferi e la industria dell'acido borico in Toscana," VI Congreso Internazionale di chimica applicata, (Rome, 1906), Atti, 1: 554-674.
A history (pp. 589-611), illustrated description of the works, and bibliography. U.F. Hoefer's memorandum of 1778, announcing the discovery (no. 644) is included as an appendix.

665. (PACIFIC COAST BORAX CO.). Borax. San Francisco, 1896. (79 pp.)
This advertising puff for a formerly scarce and now plentiful material is at once history and a primary source.

666. RAFFAELLO, Nasim. I soffioni e i laguni della Toscana e la industria borafera. Rome, 1930.
Not seen.

667. SPEARS, John R. Illustrated Sketches of Death Valley and Other Borax Deserts of the Pacific Coast. New York, 1892. (226 pp. Reprinted 1977)

Anecdotal and journalistic but an informative account of the first 30 years of borax production in California and Nevada. Promotional in intent, it appears to have predated the discovery that promotion can be accomplished without information.

668. UNITED STATES BORAX & CHEMICAL CORP. <u>Borax</u>. Los Angeles, 1960.
Give (briefly) the history and condition of the major present borax source near Boron (Mojave desert), California.

669. VER PLANCK, W.E. "History of Borax Production in the United States," <u>California Journal of Mines and Geology</u>, 52 (1956), 273-91.
A good summary history, 1864 to date. No citation but gives a chronological list of publications on borax.

CERAMICS

EARLY WORKS

670. PICCOLOPASSO, Cipriano. <u>Arte del Vassio</u>. (1548). Ed. and Eng. trans. by Berhard Rackham and Albert Van de Put. London: Victoria and Albert Museum, 1934. (85 pp., 80 plates)
Facsimile of manuscript with English translation. It deals only with earthenware.

671. DOBSON, Edward. A rudimentary treatise on the manufacture of bricks and tiles. London, 1850. (2 Pts., 119 and 103 pp.)
Reprinted in the <u>Journal of Ceramic History</u> (no. 5, 1971), where it is said to "bridge the post-medieval world with that of the mechanization of the 19th century."

HISTORIES

672. BENTON, W.E. "Evolution of Enameling," <u>Proceedings, Institute of Vitrious Enamellers</u>, 4 (1938), 29-49.
Not seen.

673. BLADEN, V.W. "The Potteries in the Industrial Revolution," <u>Economic History</u>, 1 (1926), 117-30.

Discusses economic considerations and "the technological revolution in the (British) potteries." Citations.

674. ERIKSEN, Svend. <u>Sèvres Porcelain</u>. n.p. (London ?), 1968. (339 pp.)
Originated in the mid-eighteenth century. This is mainly art history with some reference to technology.

675. FOSTER, Joseph A., ed. <u>Contributions to a Study of Brickmaking in America</u>. Claremont (New Hampshire ?): privately printed, 1962-71. (6 parts)
A miscellany of early sources, relevant English statutes from 1477 (1, 2), early accounts of brickmaking in England (3, 4), and in America (5, 6).

676. HAVARD, Francis T. "Notes on the History and Development of the Fireclay and Refractories Industries," <u>Brick and Clay Record</u>, 40 (1912), 417-19.
From Egypt to modern times. Very summary indeed but has some technical detail. No citations.

677. HEINTZ -. "Beitrag zur Geschichte der europäischen Porzellanfabrikation," <u>ZAC</u>, 5 (1898), 1156-63.
Chiefly on E.W. Tschirnhaus and J.F. Böttger. Claims that the discovery of porcelain was no accident. Few references.

678. HEINTZ -. "Geschichte der Erfindung des Porzellans durch Johann Friedrich Böttger," <u>AGNT</u>, 2 (1910), 183-200.
Based on a manuscript of Böttger, which is printed here.

679. HIPPSLEY, -. "A Sketch of the History of Ceramic Art in China," Washington, 1902 (111 pp. from <u>Annual Report of the United States National Museum</u> for 1900, 305-416).
Summary but important for its citations. It includes a catalogue of the author's collection.

680. JENKINS, Rhys. "The Silica Brick and Its Inventor, Weston Young," <u>Refractories Journal</u>, 18 (1942), 179-83, 283-90.

Dates the "first special brick" to 1696. Young (1776-1847) invented the silica brick before 1834, employing lime as a binder for "silica sand." Citations. In an article by the same title in TNS, 22 (1942), 139-47, the author dates the invention at about 1820.

681. KALKSCHMIDT, E. Der Goldmacher J.F. Böttger und die Erfindung des europäische Porzellans. Stuttgart, 6th ed., 1926. (76 pp.)
Not seen.

682. KRENKOW, F. "The Oldest Western Accounts of Chinese Porcelain," Islamic Culture, 7 (1933), 464-71.
Quotations from al-Biruni (973- c1050).

683. MARRYAT, Joseph. A History of Pottery and Porcelain, Medieval and Modern. 2nd ed., London, 1857. (472 pp.)
Brief histories of pottery industries of European regions and states are followed by accounts of porcelain in China and Europe. Citations.

684. NEWMAN, Michael. Die Deutsche Porzelan-Manufakture. Braunschweig: Klinkhardt & Bierman, 1977. (2 vols)
Principally art but has something on manufacturing at the principal German centers. Bibliography.

685. PETERS, Hermann. "Die Erfindung des Europäischen Porzelans," AGNT, 2 (1910), 399-424.
Analyzes the contribution of Tschirhaus and Böttger. Citations.

686. REIS, Heinrich, and Henry LEIGHTON. History of the Clay Working Industries in the United States. New York, 1909. (270 pp.)
A description of products with data is followed by histories of the manufacture of "clay products" in each state. References.

687. ROBERTS, Clarence N. "History of the Brick and Tile Industries of Missouri." dissertation, Univ. of Missouri, 1950. (300 pp., available from Univ. Microfilms)

The industry began about 1800 and does not appear to have been very exciting. This dissertation, however, is exemplary for its references and substantial attention to technology.

688. SHAW, Simeon. *History of the Staffordshire Potteries*. Manley, 1829. (244 pp. reprinted, Great Neck, N.Y: Weinstock, 1968)
Old fashioned but informative. No citations.

689. STRATTON, Herman J. "Factors in the Development of the American Pottery Industry, 1860-1929." dissertation, Univ. of Chicago, 1929. (360 pp., typescript in Library of Congress)
Economic history with attention to technology. Citations.

690. THOMAS, J. "The Pottery Industry and the Industrial Revolution," *Economic History*, 3 (1937), 399-414.
More on the "revolution" in the British potteries in the eighteenth century with particular reference to steam power, mechanical shaping and decoration. Intended to correct "the traditional view."

691. THOMAS, John. *The Rise of the Staffordshire Potteries*. New York: AM Kelley, 1971. (288 pp.)
Economic and social history with substantial attention to technology. Citations.

692. WASCHA, Otto. *Meissner Porzelan*. Dresden: Verlag der Kunst, 1973. (514 pp.)
Includes a substantial historical introduction.

693. WEATHERHILL, Lorna. "Technological Change and Potter's Probate Inventories, 1660-1760," *Journal of Ceramic History*, no. 3 (1970), 3-14.
Describes the evolution of raw materials, equipment, and products, on the basis of inventories of 31 potters in Burslam, Staffordshire. References.

694. WEINRICH, Peter H. *A Bibliographical Guide to Books on Ceramics*. Ottawa: Canadian Crafts Council, 1976.
Books designated "historical" are listed on pp. 17-36.

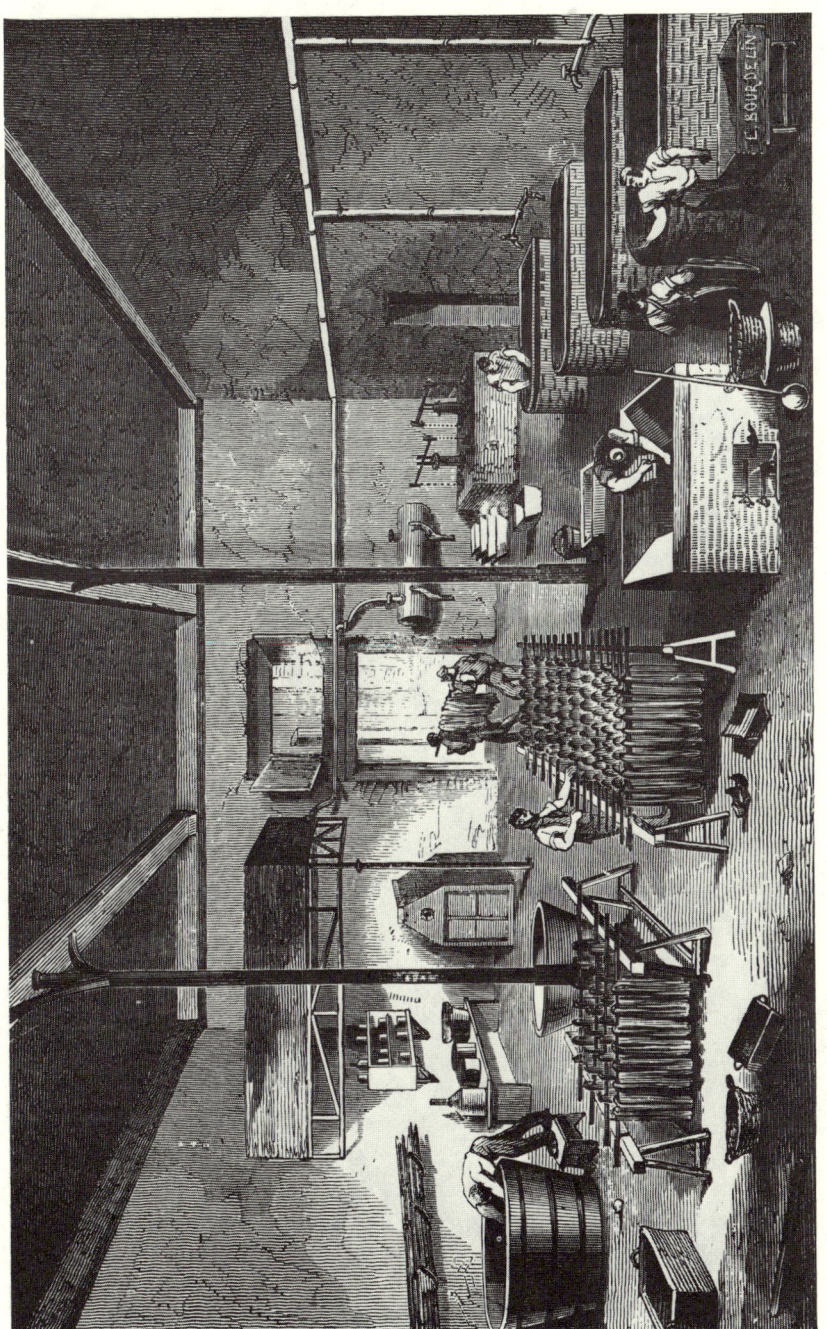

Figure 3. Interior of the dye works of Les Gobelins, France. From Turgan (item 226)

COLORS AND DYES

    EARLY WORKS

695. Compositiones ad tinguenda. ed. & German trans. by Hjalmar Hedfors, Uppsala, 1932.
Deals with the chemical and metallurgical techniques of the jeweler's art from a manuscript of the late eighth century. It seems to be the oldest of the treatises also exemplified by the Mappae clavicula, "Heraclius," and "Theophilus" (see nos. 696-98). It has been published earlier by J.M. Burman (A classical technology, Boston, 1920 [English trans.]) and by L.A. Muratori (in Antiquitates Italicae Medii Aevi, Arrettii, vol.4, 1774 [Latin]), and has been analyzed by Johnson (no. 747), who connects it to ancient tracts of the same kind.

696. Mappae clavicula. ed. & trans. by C.S. Smith and J.G. Hawthorne, Philadelphia, 1974 (Trans. APS, n.s. 65, pt. 4).
An early ninth-century work similar to and possibly contemporaneous with the Compositions ad tinguenda.

697. "HERACLIUS." De coloribus et artibus Romanorum. ed. and German trans. by Albert Ilg, as Heraclius, von den Farben und Künsten der Römer, (Vienna, 1873).
A tenth-century treatise similar to the two previous items. Also found, in English trans., in Merrifeld (no. 699), pp. 182-257.

698. "THEOPHILUS." Schedula diversarum artium. ed. and trans. by J.G. Hawthorne & C.S. Smith (as On divers arts). Chicago, 1963.
Although this work of 1130-40 is similar to the preceding items it gives more attention to basic metallurgical operations and tools of the trade. The author has been identified as a Benedictine monk, Roger of Helmarshausen.

699. MERRIFIELD, M.P. Original Treatises on the Arts of Painting. London, 1849 (reprinted N.Y.: Dover, 1967. 2 vols., continuously paginated)

A standard work of 918 pages, plus 310 pages of introduction, printing in English translation a number of medieval and Renaissance texts.

700. "The Allery Maktel (1532)." Ed., S.M. Edelstein. T&C, 5 (1964), 297-321.
Facsimile and English trans. of "the earliest printed book on spot removing and dyeing."

701. WALLER, Richard. "A catalogue of simple and mixt colours, with a specimen of each colour prefixt to its proper name," PT, 16 (1686-92), 24-32.
A "more philosophical" list of colors (14 items) than one the author reports having been printed in Stockholm in 1680. I find no "specimens" (nor any room for them).

702. "Notitia caerulei Berolensis nuper inventi. Miscellanea Berolinensis, 1 (1710), 377-78.
Refers to the discovery of Berlin (Prussian) blue.

703. WOODWARD, John. "Praeparatio caerulei Prussiaci ex Germania missa." PT, 33 (1724-25), 15-17.
A description of a process for making Prussian blue. Followed by comments by John Brown (pp. 17-24).

704. GEOFFROY, E.F. "Observations sur la preparation du bleu de Prusse, ou de Berlin," AdS Paris, 1725, 153-72, 220-37.
Cites Woodward and J.F. Henckel as earlier writers on the topic.

705. HELLOT, Jean. L'art de la teinture des laines. Paris, 1750. (631 pp., 1st ed., 1740-41)
An early work, revealing the state of the art. No history (he says that one can learn nothing from the earlier literature). An English translation appeared in 1785.

706. JUSTI, J.H.G. Das entdeckte Geheimnisz der neuen Sächischen Farben. Vienna, 1761. (1st ed. 1750)
A blue-green color perhaps a variety of Saxon blue (smalt), which is said to have been known since about 1550. Smalt was a cobalt pigment.

707. LEHMANN, J.G. Cadmiologia, oder Geschichte des Farben-Kobalds. Königsberg, 1761-66. (2 vols)
On cobalt pigments. Not seen.

708. GRAVENHORST, Die Gebruder (J.H & C.J). "Nachricht, eine grüne Mahlerfarbe, Braunschweigisch-Grün gennant, betreffend. Hannoverische Magazin, 5 (1767) 1193-1200.
Describes their discovery of "Brunswick green," a new pigment based on copper. They want to advertise it, not to reveal the process. Chemical manufacturers of Braunschweig from about 1759, the Gravenhorsts listed their products in 1771, in the same journal (pp. 130-43) to be sal ammoniac, red alum, Glaubers salt and mineral alkali "from common salt." Here again they appear to have been principally concerned with publicity. (See also no. 229.)

709. WOULFE, Peter. "Experiments to Shew the Nature of Aurum Mosaicum," PT, 61 (1771), 114-27.
Comments on recipes for this ancient yellow color and attempts to elucidate its composition (tin sulphide, also known as musivgold). The author was a London chemist and inventor of the "Woulfe bottle."

710. SCHEELE, C.W. "Zubereitung einer grünen Farbe," AdS Stockholm, 40 (1778), 316-17.
Description of "Scheele's green," a copper arsenate pigment.

711. RINMANN, Sven. "Von einer grünen Mahlerfarbe aus Kobalt," AdS Stockholm, n.s. 1, (1780), 157-66, 2 (1781), 3-12.
Description of Cobalt (or Saxon) green.

712. SCHEELE, C.W. "Ueber die farbende Materie in Berlinerblau," AdS Stockholm, n.s. 3, (1782), 256-66, 4 (1783), 32-41.
Reports the discovery of hydrocyanic acid, a landmark in the analysis of dyestuffs.

713. WESTRUMB, J.F. "Chemische Versuche über das Verhältniss des Metalls, zum färbenden Wesen, in den sogennanten Berlinerblau," (Crell's) Beiträge zu der chemische Annalen, 1 (1785-86), 42-57.

Reviews the state of research on Prussian blue. Citations.

714. BERTHOLLET, C.L. Eléméns de l'art la teinture. Paris, 1791. (2 vols)
An early standard text of sufficient repute that Andrew Ure saw fit to edit in 1891 (!) an English translation of the second edition. Begins with a 48-pp. historical introductions.

715. KAPFF, Friedrich. Beiträge zur Geschichte des Kobalts, Kobaltbergbaues, und der Blaufarbenwerke. Breslau, 1792. (160 pp.)
Not seen. Reviewed in Handlungszeitung, 10 (1793), 13-14, 22-24.

716. MERIMEE, C. "Rapport sur le prix pour la fabrication du bleu de Prusse," BSE, 2 (1803), 16-19.
Reports on the results of a prize offered in 1802, aimed at making France self-sufficient in this dyestuff.

717. CHAPTAL, J.A.C. "Observations sur les deux procédés employés pour la fabrication du verdet, ver de gris...Journal des Mines, 13 (1803), 229-32.
A chemist reports on methods of producing the traditional color, verdegris.

718. VAUQUELIN, L.W. "Sur la meilleure méthode pour décomposer le chromate de fer, obtenir l'oxide de chrome, préparer l'acide chromique, et sur quelques combinaisons de ce dernier," AC, 70 (1809), 70-94.
Includes the preparation of the pigment chrome green.

719. GMELIN, C.G. "Ueber ultramarin und dessen künstliche Darstellung..." (Erdman's) Journal für technisch- und öcomomische Chemie, 2 (1828), 379-91.
Reports the synthesis of one of the most valuable mineral pigments, ultramarine (lapis lazuli, previously obtained from an obscure central Asian source [Afghanistan]). Notes that the same invention has been claimed for (J.B.) Guimet, in response to a prize offered by the Societe d'Encouragement (Paris) in 1825 (BSE, 24 (1825), 307).

720. FRESNIUS, C.R. "Ueber Farben im technischen Sinn," NN, 8 (1858), 60-90.
Important discussion by a leading analyst of the standard dyes and mineral colors in terms of their composition and manufacture.

HISTORIES

721. "Medieval Dyeing," Ciba Review, no. 1 (1937), 3-31.
Articles in this periodical are typically authoratative, illustrated, and with a minimum of citation. Where authors are clearly identified the articles are listed under their names.

722. "India, its dyers and its color symbolism," Ciba Review, no. 2 (1937), 38-61.

723. "Scarlet," Ciba Review, no. 7 (1938), 206-32.

724. "Weaving and dyeing in ancient Egypt and Babylon," Ciba Review, no. 12 (1938), 390-417.

725. ADROSKO, R.J. Natural Dyes in the United States. Washington: Smithsonian Institution, 1968. (160 pp.)
The first part (54 pp.) gives a good account of the natural (i.e., presynthetic) dyes used in the United States--which were much the same as those used in Europe. The rest is a history of manuals and recipe books published in the United States before 1870. Bibliography.

726. BANCROFT, E. Experimental Researches Concerning the Philosophy of Permanent Colours. London, 1813 (2 vols)
Revision of a work of the same title published in 1794. Historical introduction and considerable history in the text, especially on cochineal and Prussian blue.

727. BEER, John J. "Eighteenth Century Theories on the Process of Dyeing," Isis, 51 (1960), 21-30.
On Colbert's attempt to improve dyeing in France from about 1670 and its consequences (moderate) in the work of C.F. Dufay, Jean Hellot, P.J. Macquer, Torbern Bergman, and C.L. Berthollet.

728. BISCHOF, J.N. Versuch einer Geschichte der Farber-
kunst. Stendal, 1780. (275 pp.)
Deals essentially with the "natural history" of dye
materials.

729. BORN, Wolfgang. "Purple," Ciba Review, no. 4 (1937),
106-29.

730. BRUNELLO, Franco. The Art of Dyeing in the History of
Mankind. Vicenza: Pozza, 1973 (467 pp.)
Comprehensive with references. An excellent recent
work translated from the Italian ed. of 1968.

731. CALEY, E.R. "Color in Ancient Times," in White, 1956
(no. 1262), 233-37.
A succinct and useful summary of the work of the
author and others on this topic. References.

732. CHAPTAL, J.A.C. "Sur quelques couleurs trouvés à
Pompeia," AC, 70 (1809), 22-31.
Analysis of seven samples of several colors found in
the shop of a color merchant.

733. CLARK, H.O., and Rex WAILES. "Preparation of Woad in
England," TNS, 16 (1936), 69-95.
Primarily a description as of 1930 of the processes,
machinery and costs of the last surviving woad factory
in England.

734. COLBY, L.J.M. "History of Prussian Blue," AS, 4
(1939/40), 206-11.
The pigment was discovered in 1704 and has been the
subject of scientific study ever since.

735. DAVY, Humphrey. "Some Experiments and Observations on
the Colours used in Painting by the Ancients," PT,
105 (1815), 97-124.
An analysis of fragments found by archeologists at
Rome.

736. EDELSTEIN, Sidney. Historical Notes on the Wet Processing Industry. n.p., Pt. 1 (1956), Pt. 2 (1964).
(54 & 76 pp.)
23 articles, mostly on the history of dyes and dyeing,
all previously published in the American Dyestuffs
Reporter. Citations.

737. EIBNER, Alexander. Entwicklung und Werkstoffe der Wandmalerei von Altertum bis zur Neuzeit. Munich, 1926. (618 pp.) Supplement, 1928 (195 pp.) Not seen.

738. FABER, G.A. "Dyeing and Tanning in Classical Antiquity," Ciba Review, no. 9 (1938), 278-308.

739. GERSCHEL, Lucien, "Couleur et teinture chez divers peuples indo-européens," Annales: Economies, Sociéties, Civilisations, 21 (1966), 608-31.
The symbolism of three, later four, colors, among Indo-European peoples "has already been well studied." The question here is, what were the colors? Fascinating although inconclusive. Deals with India, Greece, Rome, and medieval Europe.

740. GETTENS, R.J. "Lapis Lazuli and Ultramarine In Ancient Times," Alumni, 19 (1950), 342-57.
Concludes that most, if not all, of the mineral (lapis lazuli) and its pigment (ultramarine) came from a single location in Afghanistan. Bibliography.

741. GIBBS, F.W. "Historical Survey of the Japanning Trade," AS 7 (1951), 401-16. 9 (1953), 88-95, 197-213, 214-22.
Lacquer work from the Far East reached Europe in the sixteenth century and interested scientists as early as Boyle. At the same time attempts began to adapt the technique to metal surfaces--"japanning." The last three parts are concerned with this trade in England. An important study with citations.

742. HAUSBRAND, -. "Beiträge zur Geschichte der Blaufarbenwerke," Zeitschrift für der Berg- Hütte- und Salinenwesen, 84 (1936), 517-45.
A comprehensive history of Blaufarben (smalt, a mineral pigment based on cobalt) from 1665 to 1872 as recorded in unpublished German documents. Cites earlier literature and gives histories of individual factories.

743. HAUSMANN, J.F.L. "Beiträge zur Geschichte der Niello-Arbeit," Archiv für Mineralogie, Geognosie, Bergbau und Hüttenkunde (Berlin), 23 (1850), 432-43.

A survey of literary references to the metal-coloring technique called niello.

744. HEUSSER, A.H., ed. The History of the Silk Dyeing Industry in the United States. Patterson, N.J., 1927. (602 pp.)
Sponsored and published by the Silk Dyers Assn. of America. Although strongly biographical the book gives considerable attention to background and is altogether more substantial than one would expect.

745. HURRY, J.B. The Woad Plant and Its Dye. Oxford, 1930. (328 pp.)
Comprehensive with much history.

746. HUTH, Hans. Lacquer of the West: the History of a Craft and an Industry, 1550-1950. Chicago: Univ. of Chicago Press, 1971. (158 pp., 364 plates. trans. of his Europäische Lackarbeiten.)
He dates the "industry" from the last half of the seventeenth century at Spa (Belgium). An elegant book primarily concerned with art. Citations.

747. JOHNSON, Rozelle P. Compositiones variae: an introductory study. Urbana, Ill., 1939. (116 pp.)
An analysis of the Compositiones ad tinguenda (no. 695), and comparison with similar works.

748. KASWELL, E.R. "Color Designation through the Ages," in White, 1956 (no. 1262), 239-47.
An important summary. Citations.

749. KING. P.E. "The present state of Development of the Theory of Dyeing," Journal, Society of Dyers and Colourists, 35 (1919), 171-77.
Noteworthy for its meticulous citation of sources from the early eighteenth century to date of publication.

750. LAUTERBACH, Fritz. Geschichte der in Deutschland bei der Färberei angewandten Farbstoffe, mit besonderer Besuchtingung des mittelalterlichen Waid Bases. Leipzig, 1905.
Not seen.

751. LAWRIE, L.G. A Bibliography of Dyeing and Textile
     Printing, Comprising a List of Books from the 16th
     Century to the Present Time. London, 1949.
     Alphabetical list of 804 books followed by a chronological list of the same.

752. LEGETT, Wm. F. Ancient and Medieval Dyes. Brooklyn,
     1944. (102 pp.)
     A useful summary. No citations.

753. LEIX, Alfred. "Trade Routes and Dye Markets in the
     Middle Ages," Ciba Review, no. 10 (1938), 314-40.

754. MELLOR, C.M., and D.S.L. CARDWELL. "Dyes and Dyeing,
     1775-1860," British Journal of the History of
     Science, 1 (1962-63), 265-79.
     A description of British practice, based principally
     on the records of two firms at Leeds. Concludes that
     "the dyeing industry seems to have been as scientific
     as one could have wished at that period." Citations.

755. MINUTOLI, Heinrich. "Ueber die Pigmente und die Malertechnik der Alten, besonders der alten Aegypter,"
     (Erdman's) Journal für technische und ökonomische
     Chemie, 8 (1830), 173-87.
     Primarily a bibliographical survey.

756. NOBEL, Joseph V. "The Technique of Egyptian Faïence,"
     American Journal of Archeology, 73 (1969), 435-39.
     "Egyptian faïence, the earliest glazed ceramic material, was produced (from end 5th C B.C.) by means of a
     single self-glazing formula making use of natron..."
     Includes analyses and experimental formulations.
     Citations.

757. NUNGESSER, Thilo. "The Dyeing of Leather: Outlines of
     the History and Present Day Development," Die BASF,
     20 (1970), 125-32.
     Summary. No references.

758. PARTINGTON, J.R. "The Discovery of Mosaic Gold," Isis,
     21 (1934), 203-06.
     Calls attention to a 14th century Italian manuscript
     in Naples which he suggests may contain the earliest
     reference to this "purple color," tin disulphide, made
     from tin amalgam, sulphur, and sal ammoniac.

759. PLOSS, E.E. Ein Buch von alten Farben. Heidelberg & Berlin, 1962. (188 pp.)
An elegantly illustrated and well-documented history of dyeing during the era of natural dyes.

760. PRANGE, C.F. Farbenlexikon. Halle, 1782. (572 pp.)
Not seen.

761. ROBINSON, Stuart. A History of Dyed Textiles. London, 1969. (122 pp.)
Summary but comprehensive, without citations but with a copious bibliography. Up to date.

762. SACC, -. "Esquisse de l'histoire de la pourpre," Bull. de la Société Industrielle de Mulhouse, 26 (1854), 305-08.
A chronology of events in European research on purple colors from 1683.

763. SCHMAUDERER, E. "Kenntnisse über das Ultramarin bis zur ersten künstlichen Darstellung um 1827," Technikgeschichte, 36 (1969), 147-60.
This is one of a set of articles, which also includes his "Die Entwicklung der Ultramarin-Fabrikation im 19. Jahrhundert," Tradition, 3-4 (1969), 127-52, and "Künstliches Ultramarin im Spiegel von Preisaufgaben und der Entwicklung der Mineralanalyse im 19. Jahrhundert," Technikgeschichte, 36 (1969), 314-33. Numerous citations in all.

764. SCHULZ, R. Aus der Geschichte des Farbstoffhandels in Mittelalter. Munich, 1929. (129 pp.)
A very useful dissertation with citations.

765. THOMAS, Parakunnel J. "The Beginnings of Calico Printing in England," English Historical Review, 39 (1924), 207-16.
The first factory opened in 1690, arousing competition among the wool and silk weavers which led to legislation banning the wearing of cotton cloth, a ban which extended from 1720 to 1774. Based on Board of Trade papers in the Public Record Office (London).

766. THOMPSON, Daniel V. "Artificial Vermillion in the Middle Ages," Technical Studies in the Fine Arts, 2 (1933), 62-70.

Natural vermillion was mercuric sulphide, as "cinnabar," the ore of mercury. Artificial vermillion was the same, made by reacting mercury with sulphur. He calls it the most important new pigment of the period.

767. THOMPSON, Daniel V. The Materials And Techniques of Medieval Painting. London, 1936. (239 pp., reprinted, New York: Dover)
A standard work.

768. TURNBULL, Geoffrey. A History of the Calico Printing Industry of Great Britain. Altringham, 1951. (500 pp.)
Some citations in the text. Appends a chronological list of literature on bleaching, printing and dyeing, from Pliny the Elder to 1859.

769. VETTERLI, A. "The History of Indigo," Ciba Review, no. 85 (1951), 3066-96.

770. WEHLTE, K. The Materials and Techniques of Painting. New York, 1975. (678 pp. trans. of hi Werkstoffe und Techniken der Malerei, Ravensburg, 1967)
A modern textbook but the systemic treatment of each pigment includes a valuable paragraph on its history, many with citations.

771. WESCHER, H. "Dyeing in France before Colbert," Ciba Review, no. 18 (1939), 618-25.

772. WINKLER, August F. Das sächsische Blaufarbenwesen von 1790 in Bilde. Berlin, 1959. (27 pp. Freiberger Forschungshefte D 25)
Not seen.

FOOD & DRINK (see also sugar)

EARLY WORKS

773. APPERT, Nicholas. L'art de conserver, pendant plusieurs années, toutes les substances animales et vegetales. Paris, 1811. (116 pp.)
The pioneer work on "canning." Numerous later editions, translations, and reprints.

HISTORIES

774. Index to the Literature of Food Investigation, No. 1, London: HMSO, 1929.
An historical review of literature on food preservation by artificial cold covers, pp. 1-8.

775. "Historic tinned foods," Chemistry and Industry, 57 (1938), 300-14, 327-36, 914-17. (also published separately as pub. no. 85 of the Tin Research and Development Council (London), 1935. (70 pp.)
After an historical introduction, consists of scientific descriptions of surviving preserved foods, from 1771 ("a cake of portable soup") to the late nineteenth century.

776. ALLEN, H. Warner. A History of Wine: Great Vintage Wines from the Homeric Age to the Present Day. London, 1961. (304 pp.)
Deals only with "fine old vintage wines." Subdivided into the first golden age (Homer to Galen [200 AD.]), the "dark age" (to about 1600), the second golden age (to about 1900), and the present "silver" age. Citations only for the first golden age!

777. ANDERSON, Oscar E. Refrigeration in America. Princeton, 1953. (344 pp.)
Deals with all aspects of the question. Thorough and scholarly. Citations.

778. ANDERSON, Oscar E. The Health of a Nation: Harvey W. Wiley and the Fight for Pure Food. Chicago, 1958. (332 pp.)
A biography, giving extensive detail on the background of the United States Pure Food and Drug Act of 1906. Citations.

779. ARNOLD, John P. Origin and History of Beer Brewing. Chicago, 1911. (411 pp.)
Comprehensive with many citations to and quotations from primary sources. Numerous illustrations.

780. ARNTZ, Helmut. Weinbrenner: die Geschichte von Geistes des Weines. Stuttgart, 1975. (284 pp.)

A somewhat strangely organized but scholarly and fascinating account of the history of distilled beverages up to 1900. Gives particular attention to mystical and alchemical aspects of the subject. Citations.

781. BARON, S.W. Brewed in America: A History of Beer and Ale in the United States. Boston, 1962. (424 pp.)
Primarily social and economic history but well documented and with an extensive bibliography. Gives some attention to the European background.

782. BIRDSEYE, Clarence, and G.A. FITZGERALD, "History and Present Importance of Quick Freezing," I&EC, 24 (1932), 676-78.
A good summary with citations. The technique became commercially significant with Birdseye's patents of 1930.

783. BITTING, A.W. Appertizing, or the Art of Canning: Its History and Development. San Francisco, 1937. (852 pp.)
"It is the object of this work to present the more important facts in the history and development..." With these words from the preface the reader is plunged, without a table of contents, into a jungle of material, much of it historical. As a guide he has the index, which begins, Ainsley, J.C., pioneered canning fruits for salad; Appert, portrait: Appert, title page; Appert's work, Edinburgh Review; apple; apple, baked; apple butter; and so on (with eight more entries for "apple").

784. BROWN, H.T. "Reminiscences of Fifty Years' Experience of the Application of Scientific Method to Brewing Practice," Journal, Institute of Brewing, 22 (1916), 267-353.
Not seen.

785. CLACQUESIN, Paul. Histoire de la communauté des distillateurs: histoire des liquors. Paris, 1900. (329 pp.)
Not seen.

786. COPPOCK, John B.M. "The Evolution of Food Science and Technology in the United Kingdom," Chemistry and In-

dustry, (1973), 455-60.
An informative summary with references.

787. CORRAN, H.S. *A History of Brewing*. Newton Abbot: David & Charles, 1975. (303 pp.)
A comprehensive and systematic history with references in the text but largely restricted to Britain.

788. CUTTING, C.L. *Fish Saving: A History of Fish Processing from Ancient to Modern Times*. London, 1955. (372 pp.)
Comprehensive with citations. An important work.

789. DISANTI, Evelyn, "Giant Steps for Food Technology," *Food Technology*, 31 (1977), May 54-60.
Describes foods developed for the National Aeronautics and Space Administration (U.S.) over two decades.

790. DOWNARD, Wm. L. *Dictionary of the History of American Brewing and Distilling Industries*. Westport, Conn.: Greenwood, 1980 (268 pp.)
A dictionary of terms and (mostly) firms.

791. EMERSON, Edward R. *Beverages, Past and Present: An Historical Sketch of Their Production, Together with a Study of the Customs Connected with Their Use*. New York, 1908. (2 vols)
A chatty tour around the world from Japan to North America and not missing much. Quotations but no references.

792. ENJALBERT, Henri. *Histoire de la vigne et du vin*. Paris: Bordas, 1975. (207 pp.)
Especially important for treatment of the (seventeenth and eighteenth century) "revolution in drinks," and for a history of the "contemporary era." No references.

793. EYER, Fritz. "Brauer und Brauen im alten Frankreich," *GGBB*, 1973, 9-28.
Citations.

794. FILBY, F.A. *A History of Food Adulteration and Analysis*. London, 1934. (265 pp.)
A good circumstantial history, although brief, with citations and numerous appendices.

795. FORBES, R.J. "The Rise of Food Technology, 1500-1900," Janus, 47 (1958), 101-27, 139-55.
And excellent summary account with bibliography.

796. FRANCIS, Clarence. A History of Food and Its Preservation. Princeton, 1937. (45 pp.)
Not seen.

797. FREY, Charles N. "History and Development of the Modern Yeast Industry," I&EC, 22 (1930), 1154-62.
A useful summary with references.

798. GESELLSCHAFT FUER DIE GESCHICHTE UND BIBLIOGRAPHIE DES BRAUWESENS. Jahrbuch. Berlin, 1928 ff.
In addition to annual bibliographies, it contains articles, many of high quality, on all aspects of brewing, mostly but not exclusively with reference to Germany. Some articles of general interest have been entered separately (abbreviated GGBB).

799. GIEDION, Siegfried. Mechanization Takes Command. New York, 1948. (743 pp.)
This celebrated critique of modern technology and its impact gives substantial attention to the food industries, especially flour, bread, and meat.

800. GOLDBLITH, Samuel A. "Historical Development of Food Irradiation," International Symposium on Food Irradiation. Proceedings, 1 (1966), 3-17.

801. GOTTSCHALK, Alfred. Histoire de l'alimentation et de la gastronomie. Paris, 1948. (2 vols)
Although there isn't much about technology, this is a very comprehensive history of food--although in modern times referring particularly to France (and why not?). Some references in text.

802. GREAT BRITAIN, Committee on food preservation. Report...inquiring into the Use of Preservatives and Colouring Matters in the Preservation and Cooking of Food. London: HMSO, 1901.
Report of hearings held in 1899. Much historical information.

803. HARTMANN, L.F., and A.L. OPPENHEIM, "Beer and Brewing Technologies in Ancient Mesopotamia," Journal of the American Oriental Society, Supplement, no. 10, 1950.
Not seen.

804. HELCK, Wolgang, "Das Bier in alten Aegypten," GGBB, 1972, 9-120.
Citations.

805. HENDERSON, Alexander. The History of Ancient and Modern Wines. London, 1824. (408 pp.)
A belated attempt to fulfill Francis Bacon's demand for a history of wine. A systematic treatment of ancient wine is followed by a similar history of modern wines, country by country. Citations.

806. INTERNATIONAL ASSOCIATION OF ICE CREAM MANUFACTURERS. The History of Ice Cream. Washington, 1951. (43 pp.)
Not seen.

807. JACOB, H.E. Six Thousand Years of Bread. New York, 1944. (399 pp.)
Popular and mostly anecdotal without references.

808. KEUCHEL, Edward F. "Science, Technology and Food Preservation: The Introduction of Bacteriology to the American Canning Industry," in Burton J. Williams, ed., Essays in American History in Honor of James C. Malin. Lawrence, Kansas: Coronado Press, 1973, 163-78.
The Wisconsin Agricultural Experiment Station applied bacteriology to canning in 1894.

809. KIRKBY, William. The Evolution of Artificial Mineral Waters. Manchester, 1902. (155 pp.)
Informative, especially on apparatus. Citations.

810. LEVEY, Martin, "Food and Its Technology in Ancient Mesopotamia: The Earliest Chemical Processes and Chemicals," Centaurus, 6 (1959), 36-51.
Refers briefly to all aspects of the subject with citations (169!) to the literature of cuneiform studies.

811. LIPPMANN, E.O. von. "Zur Geschichte der Konserven und des Fleischextraktes," in Lippmann, Abhandlung (no. 129) vol. 1:343-46. (from Chemiker Zeitung, 1899)
A brief note mentioning several predecessors of Appert.

812. LIPPMANN, E.O. von. "Geistige Getränke im fruhmittelalterlichen Indien," In Lippmann, Beiträge (no. 130) vol. 2:204-09. (from Chemiker Zeitung, 1927)
Citations.

813. MATHIAS, Peter. The Brewing Industry in England. Cambridge, 1959. (595 pp.)
An immense and well-documented work, in which, however, technology is only briefly treated.

814. MAURIZIO, Adam. Geschichte der gegorenen Getränke. Berlin, 1933. (262 pp.)
A comprehensive and scholarly history of fermented beverages. Citations.

815. NIXON, H.C. "The Rise of the American Cottonseed Oil Industry," Journal of Political Economy, 38 (1930), 73-85.
Reports that cottonseed was garbage in 1860, fertilizer in 1870, cattle feed in 1880, and table food in 1890. References.

816. OLBRICH, Hubert. Geschichte der Melasse. Berlin: Institut für Zuckerindustrie, 1970. (831 pp.)
Everything you wanted to know about molasses. Citations, bibliography, and a chronology beginning in 350-375 A.D.

817. RAOUL, Yves. "Vitamins, Their History and Their Research," Die BASF, 20 (1970), 174-81.
Interesting summary. No references.

818. RILEY, John J. A History of the American Soft Drink, Industry, 1807-1957. Washington, 1958. (302 pp.)
Includes a chapter on early nineteenth-century developments in Europe. Gives appropriate attention to technology. Citations.

819. ROELLIG, Wolfgang. "Das Bier im alten Mesopotamien," GGBB, 1971: 9-104.
Citations.

820. SCHUETTE, H.A. "Know Your Foods--I. Oleomargarine," JCE 5 (1928), 1621-26.
Summary without citations. Attributes the invention to Mége Mauries, a Frenchman, holder of a United States patent of 1873.

821. SCHUETTE, H.A., and F.J. ROBINSON, "Ice Cream," JCE, 10 (1933), 469-75.
Summary history, with emphasis on the United States.

822. SHARRER, G. Terry. 1001 References for the History of American Food Technology. Davis, Calif.: Univ. of California, Agricultural History Center, 1978. (103 pp.)
Very useful. Mimeograph.

823. SLOAN, A. Elizabeth, ed. "Symposium: 200 Years of Food: A Historical Perspective," Food Technology, 30 (1976) 31 et seq. (19 pages, some unnumbered)
Contains articles on sun drying, frozen food and American food laws, too brief (much space given to "the future") to be very useful. Citations.

824. STORCK, J., and W.D. TEAGUE. Flour for Man's Bread, a History of Milling. Minneapolis, 1952. (388 pp.)
Comprehensive although restricted to the United States in its last half. Originally written as a narrative for a museum which was never built. No citations.

825. STUYVENBERG, J.H. van. Margarine. An Economic, Social, and Scientific History. Toronto, 1969. (342 pp.)
Comprehensive. Citations.

GLASS

EARLY WORKS

826. NERI, Antonio. L'arte vetraria, Florence, 1612. (120 pp.)
Neri was a Florentine priest who worked in glass at Florence, Pisa and Antwerp. Several Italian editions were followed by an English translation by Christopher Merrett (1662), with additions expanding the book to 362 pp. All subsequent editions were based on Mer-

ritt, including a German version of 1679 with further additions by Johann Kunckel (472 pp.), and an unacknowledged translation into French by Haudiquer de Blancourt (1697). Although glass-making had been briefly treated before, this is the first book on the subject.

HISTORIES

827. (BACON, J.B.F.) "Captain John B. Ford Founded Plate Glass Manufacture in America," <u>Bulletin, American Ceramic Society</u>, 18 (1939), 263-67.
Ford (1811-1903) made plate glass in 1870 and established ten years later what was to become the Pittsburgh Plate Glass Co. An imaginative but circumstantial biography, with no indications of sources.

828. BONDOIS, P.M. "La Développement de la verrerie française au 18e siècle," <u>Revue d'Histoire Economique et Sociale</u>, 23 (1936-37), 237-61.
Coal began to replace wood as fuel from 1725 and after 1760 competition forced the abandonment of other ancient practices. Primarily an economic study. Many citations to archival sources.

829. BUSSOLIN, Dominique. <u>Les célèbres verreries de Venise et de Murano: description historique, technologique et statistique...</u> Venice, 1846. (88 pp.)
Not seen.

830. CAYLEY, E.R. <u>Analysis of Ancient Glasses, 1790-1957.</u> Corning, N.Y.: Corning Glass Museum, 1962. (118 pp.)
Records and discusses 500 analyses of glass and its raw materials.

831. CHAMBON, Raymond. <u>L'Histoire de la verrerie en Belgique du IIe siècle à nos jours.</u> Brussels: Librarie Encyclopédique, 1955. (331 pp.)
Comprehensive with numerous citations.

832. CHANCE, W.H.S. "The Optical Glass Works at Benediktbeuern," <u>Proceedings of the Physical Society</u> (London), 49 (1937) 433-43.
A summary account from German sources of Joseph Fraunhofer's glass works.

833. CHARLESTON, Robt. J. "Glass Furnaces through the Ages," Journal of Glass Studies, 20 (1978) 9-33.
From the fourteenth century B.C. (18th Egyptian dynasty) to the eighteenth century A.D. Differentiates "northern and "southern" kiln types. Citations.

834. COCHIN, Augustin. La manufacture de glaces de St. Gobain de 1665 à 1885. Paris, 1866. (192 pp.)
Not seen. Frémy (no. 841), who used it as a source, said that it was not "mis dans le commerce." He describes it as having 93 pages of history followed by 24 appendices "full of interesting anecdotes."

835. DAVIS, Pearce. The Development of the American Glass Industry. Cambridge, Mass.: Harvard Univ. Press, 1949. (316 pp.)
Economic history with considerable attention to technology. Citations and important bibliography.

836. DOUGLAS, R.W., and Susan FRANK. A History of Glassmaking. Henly on Thames:Foulis, 1972. (218 pp.)
Intended as "a convenient summary," which it is, with numerous illustrations. Bibliography but no citations. Up to date.

837. DUNCAN, G.S. Bibliography of Glass. London:Dawsons, 1960. (544 pp.)
Over 16,000 entries to 1940. Definitive.

838. EISEN, G.A. Glass: Its Origin, History. New York, 1927. (2 vols)
Mostly on art glass.

839. ENGLE, Anita, ed. Readings in Glass History. Jerusalem: Phoenix, 1973 ff.
An important serial, no. 12 of which appeared in 1981. Topics treated include glassmaking in the Medieval Near East, in Lorraine, Elizabethan England, Sidon, Altare (Italy), Jewish glassmakers, glass and the silk route, and other topics. Bibliographies and references in text.

840. FORSING, P. Glass Vessels before Glass-blowing. Copenhagen: Munksgaard, 1940. (168 pp.)
Not seen.

841. FREMY, Elphège. <u>Histoire de la manufacture royale des glaces de France au 17ᵉ et au 18ᵉ siècle.</u> Paris, 1909. (444 pp.)
Comprehensive with numerous citations to archival sources.

842. GASPARETTO, A. <u>Il vetro di Murano</u>. Venice:Neri Pozza, 1958. (287 pp.)
An account of the origin of Murano glass and the establishment of the industry there occupy the first 73 pp. Numerous citations.

843. GODFREY, Elanor S. <u>The Development of English Glassmaking, 1560-1640</u>. Chapel Hill, N.C.: Univ. of North Carolina Press, 1975. (288 pp.)
Primarily an economic history but deals with technology to test John Nef's conclusions (no. 295) concerning the importance of the introduction of coal as fuel. Concludes that the immigrant glassmaker was of primary importance, the use of coal second. During this time window glass replaced luxury tableware as the most important product. Bibliography and numerous citations.

844. HALAHAN, B.C. "Chiddingfold Glass and its Makers in the Middle Ages," <u>TNS</u>, 5 (1924), 77-85.
Glass was made from the mid-thirteenth century at Chiddingfold, Surrey, first by a family of French origin.

845. HATCH, Chas. E., Jr. "Glass-making in Virginia, 1607-1625," <u>William and Mary Quarterly</u>, 21 (1941), 119-38, 227-38.
At Jamestown on a site discovered in 1931. Citations.

846. HAYNES, E.B. <u>Glass through the Ages</u>. London: Penguin, 1948. (240 pp.)
Summary account with emphasis on art glass and largely restricted to England. Brief bibliography, no citations.

847. HORN, Georg. <u>Die Geschichte der Glasindustrie und ihrer Arbeiter.</u> Stuttgart, 1903. (368 pp.)
A "social study," primarily of labor but with significant attention to technology. Citations.

848. HOVESTADT, Heinrich. *Jena Glass*. London, 1902.
History is the concern of pp. 1-22.

849. HULME, E.W. "On the Invention of English Flint Glass,"
TNS, 6 (1925-26), 75-84.
A lead glass, discovered in "the East" but industrialized in England at the end of the seventeenth century. A slight article, supplementing his own earlier writings in *The Antiquary*, 34 (1898).

850. ILG, Albert, and Wenddin BOEHIM. *Die Glasindustrie, ihre Geschichte*... Stuttgart, 1874. (324 pp.)
There is an extended compilation of worldwide statistics on the industry as of about 1870.

851. JOURNAL OF GLASS STUDIES. 1959 ff.
An annual volume published by the Corning Glass Museum, Corning, New York. A commendable blending of pictorial elegance and meticulous scholarship, it is primarily devoted to art glass but not exclusively. Includes an annual bibliography of recent literature.

852. KAMPFER, Fritz, and K.G. BEYER. *Glass: a World History*. Greenwich, Conn.: New York Graphic Society, 1967.
Translated from German. A picture book the principal historical value of which is in the references to literature on specific items illustrated.

853. KENYON, G.H. *The Glass Industry of the Weald*. Leicester: University Press, 1967. (231 pp.)
A profound archeological study with full attention to technology. Citations.

854. KISA, Anton. *Das Glas in Altertum*. Leipzig, 1908.
(3 vols)
A comprehensive description and analysis of glass objects revealed by archeology.

855. KNOWLES, J.A. "The History of Copper Ruby Glass," *TNS*, 6 (1925-26), 66-74.
Colored by finely divided copper. He finds the art in "Theophilis" (twelfth century, no. 689), but says that it was lost by the seventeenth century and rediscovered in the twentieth. Citations.

856. LARNDER, Dionysius. A Treatise on the Origin, Progressive Improvement, and Present State of the Manufacture of Porcelain and Glass. London, 1832. (334 pp., reprinted Park Ridge, N.J.: Noyes, 1972)
Each section begins with an undocumented historical introduction.

857. LERNER, Franz. Geschichte des deutschen Glaserhandwerks. Stuttgart, 1950. (145 pp.)
"Ein Ueberblick" but a good one, of the glass industry as a handicraft. Bibliography, no citations.

858. LIDDEL, W.A. "The Development of Science in the American Glass Industry 1880-1940. diss., Yale Univ., 1970. (Univ. Microfilm, 70-2840)
Not seen.

859. MODES, C.H. "How the Old Timers Made Glass," Ceramic Industries, 35 (1940, Sept.), 62, 64.
Prints six recipes for green and clear bottle glass, of unknown origin but which the author estimates to be sixteenth to eighteenth century.

860. NEUBURG, Frederic. Ancient Glass. Toronto: Univ, of Toronto Press, 1962. (110 pp.)
Ranges in time and space from Babylonia to Byzantium, finding that glassmaking began in Egypt. Trans. from German ed. of 1962.

861. OPPENHEIM, A.L., et al. Glass and Glassmaking in Ancient Mesopotamia. Corning, N.Y.: Corning Glass Museum, 1970. (242 pp.)
"An edition of the cuneiform texts which contain instruction for glassmakers, with a catalog of surviving objects."

862. PELIGOT, Eugene. Le Verre: son histoire; sa fabrication. Paris, 1877. (489 pp.)
Despite its promise this contains little history.

863. PINCHART, A. "Les fabriques de verres de Venise, d'Anvers, et de Bruxelles au xvie et xviie siècles," Bulletin des Commissions Royales d'Art et d'Archéologie (Brussels), 21 & 22 (1882-83). (65 pp.)
Not seen.

864. POWELL, Harry. Glass-making in England. Cambridge, 1923. (183 pp.)
From Roman times to 1918. A history of the handicraft of glassmaking (as distinguished from "the craft of the collector" and "the craft of the dealer in antiques"). No systematic citation and no bibliography but includes a list of citations from 1567 to 1700, "extracted" from Parliamentary papers and contemporary newspapers.

865. PRIS, C. "La manufacture des glaces de Saint-Gobain avant la révolution industrièlle," Revue d'Histoire Economique et Sociale, 52 (1974), 161-72.
Not seen.

866. RADEMACHER, F. Die deutscher gläser des Mittelalters. Berlin, 1933. (151 pp.)
Not seen.

867. SCHEBEK, E. Böhmens Glasindustrie und Glashandel. Quellen zu ihrer Geschichte. Prague, 1878. (343 pp.)
Not seen.

868. SCHOFIELD, R.E. "Josiah Wedgewood and the Technology of Glass Manufacture," T&C, 3 (1962), 285-97.
Prints a tract written by Wedgewood in 1783 on the cause of defects in flint glass. Connects the subsequent decline of British leadership in the manufacture of optical glass with the neglect of this treatise.

869. SCHULZ, Hans. Die Geschichte der Glaserzeugung. Leipzig, 1928. (130 pp.)
Not seen.

870. SCOVILLE, Warren C. "Growth of the American Glass Industry to 1880," Journal of Political Economy, 52 (1944), 193-216, 340-55.
Citations.

871. SCOVILLE, Warren C. Revolution in Glassmaking: Entrepreneurship and Technology in the American Industry, 1880-1920. Cambridge, Mass., 1948. (398 pp.)

A case-history of the roles of economics and technology in the transformation of a "handicraft industry." Citations.

872. SCOVILLE, Warren C. <u>Capitalism and French Glassmaking, 1640-1789</u>. Berkeley, Calif., 1950. (210 pp.)
Concerned with the rise of large-scale capitalistic production. Includes a chapter on "technological innovation" which describes the difficulties of introducing coal firing after the English model. Citations and important bibliography.

873. STEUBEN GLASS CO. <u>Books on Glass</u>. New York: Steuben Glass Co., 1942. (23 pp.)
Eighty selected titles.

874. THORPE, William H. <u>A History of English and Irish Glass</u>. London, 1929. (2 vols)
Although the author excludes bottle glass and plate glass as "work, not art," his book actually contains one of the best historical accounts of the technology of glass, written in a felicitous style with extensive source citation and an important bibliography.

875. TURRIERE, Emil. "Introduction à l'histoire de l'optique: le développement de l'industrie verrière d'art depuis l'époque Venetienne jusqu'à la foundation des verreries d'optique." <u>Isis</u>, 7 (1925), 77-104.
Intended as a background to the history of optical instruments. Citations.

876. ZUMAN, Franz. "Die böhmischen Glashütten und Industrie der Glaskompositionsstein." <u>BGTI</u>, 19 (1929) 54-60.
An important summary with references from fourteenth century origins to about 1850.

LEATHER

HISTORIES

877. BASF, Leather Laboratory. <u>Leather Techniques through the Ages</u>. Ludwigshaven, n.d. (109 pp.)
Brief (mostly pictures) but informative. Citations.

878. BRAVO, G.A., and Juliana TROPKE. <u>100,000 Jahre Leder</u>. Basel, 1970. (391 pp.)

Both the most recent and the most comprehensive history I've found. Bibliography but no citations.

879. KOERNER, Theodore, "Geschichte der Gerberei," in Wolfgang Grassmann, Handbuch der Gerbereichemie un Lederfabrikation (Vienna, 1944), vol. 1: 1-89.
An excellent summary history, with citations.

880. McDERMOTT, Chas. H. A History of the Shoe and Leather Industries of the United States. Boston, 1918. (399 pp.)
Although typical of works of the time in featuring the lives of entrepreneurs, this gives substantial attention to the development of the technology. No references.

881. WELSH, P.C. Tanning in the United States to 1850: A Brief History. Washington: Smithsonian Institution, 1964. (99 pp.)
"A brief history." Straightforward and circumstantial. Numerous citations.

LIME, GYPSUM, and CEMENT

EARLY WORKS

882. CATO, Marcus Porcius (d. 149 B.C.) De agricultura. Eng. trans., New York, 1933 (reprinted 1966)
Bk. 38 describes the making of a lime kiln.

883. CEDERHIELEN, C.W. "Erinnerung von Verbesserung der Kalkbrennerey," AdS Stockholm, 1 (1739-40), 247-49.
Says that Sweden has many kinds of lime, "each having its special working, oven, and firing." But the Swedes still haven't perfected the art, and the Academy is seeking descriptions of the practices of the leading lime-burners. This is followed by one example, from the Rhineland.

884. "Observations sur la chaux," Journal Oeconomique, (1757, Jan.), 117-19.
Meticulous instructions for lime-burning, accompanied by strange excursions into theory.

885. FOURCROY DE RAMACOURT, C.R. L'art du chauloumnier.
Paris, 1766 (also included in the Description des
Arts et Métiers [no. 13], pp. 323-400 in vol. 4 of
the Neuchatel ed.)
Says there are two basic methods of lime-burning, in
one of which a wood fire is burned beneath the lime
stone; in the other intermediate layers of stone and
fuel (including coal) are ignited. Describes a variety of furnaces used in France in the former process.

886. CRUTZNACH, L.S. von. "Abhandlung von dem Kalke,"
Allgemeine Magazin der Natur, Kunst und Wissenschaft,
12 (1767), 252-80. (from Verhandeling der Weetenschap te Haarlem, vol. 5)
Notable for its citations of earlier sources. It is
his own opinion that lime is composed of earth, salt
and sulphur.

887. LANGSDORF, K.C. Entwurf zu Vorlesungen über mehrere
den Kameralisten und Technologen wichtige Gegenstände. Altenberg, 1798. (215 pp.)
Chapters on materials of construction are prominent--
lime, mortar, cement, bricks, gypsum and binding materials for fireproof construction. Good bibliographies.

888. ANDERSON, James. An Essay on Quicklime, as a Cement
and a Manure. Boston, 1799.
Rational in modern terms, mainly because he avoids
theory. His main interest is in fertilizer.

889. DEBLINNE, -, and - DONOP. "Recherches sur la forme la
plus avantageuse à donner aux fours a chaux," BSE, 9
(1810), 287-92.
Submitted in response to a prize question.

HISTORIES

890. DAVIS, A.C. A Hundred Years of Portland Cement, 1824-
1924. London, 1924. (281 pp.)
Narrative, largely restricted to England, with citations in text. Claims that Joseph Aspedin of Leeds
originated the name and perhaps the material in 1824.
Appendix of important patents.

891. FRANCIS, A.J. The Cement Industry, 1796-1914: A History. Newton Abbott: David & Charles, 1977.
A history of the British cement industry to 1900, giving substantial attention to the period of "Roman cement," before the invention of Portland cement.

892. GESSNER, A. "Römischer Kalkbrennofen bei Brugg," Anzeiger für schweizischer Altertumskunde, n.s. 9 (1907) 613 pp. (with a plate).
Drawing and description of a Roman lime kiln discovered in mining operations.

893. HALSTEAD, P.E. "The Early History of Portland Cement," TNS, 34 (1961-62), 37-46.
Says the development was "evolutionary." The paper focuses on "technical aspects," rather than on persons. References.

894. LESLEY, Robt. W. History of the Portland Cement Industry in the United States. Chicago, 1924. (330 pp.)
A detailed narrative with much quotation from documents. The same author had published in 1900 a book with the same title (Philadelphia: American Cement Co., 145 pp.) which was an interesting but rather chaotic collection of illustrations and documents.

895. LEVEY, Martin. "Gypsum, salt and soda in ancient Mesopotamia," Isis, 49 (1958), 336-41.
Cites evidence of the use of gypsum in plaster and in medicine.

896. PASLEY, C.W. Observations on Limes, Calcareous Cements, Mortars, Stuccos and Concrete. London, 1838. (288 pp.)
A pioneering work, summarizing the earlier literature.

897. QUIETMEYER, Friedrich. Zur Geschichte der Erfindung des Portlandzementes. Berlin, 1912.
Not seen.

898. ROHLAND, -. "Aus der Geschichte der Mortelmaterialen," AGNT, 2 (1910), 91-96.
Gives an analysis of mortar used in the pyramid of Cheops (gypsum and lime). Summarizes later compositions to the nineteenth century. No references.

899. ROSSMAN, J. "Patent History of Lime," <u>Rock Products</u>, 32 (1929) 66-72.
In the United States.

900. SCHARROO, P.W. "L'invention du ciment Portland," <u>AI</u>, 11 (1958), 270-74.
Argues that the "true" inventor was Isaac C. Johnson in 1844.

901. SKEMPTON, A.W. "Portland Cements, 1843-87," <u>TNS</u>, 35 (1962-63), 117-52.
On the development of high-strength concrete, with a bibliography of earlier histories and of primary sources.

902. SPACKMAN, Charles. <u>Some Writers on Lime and Cement from Cato to the Present Day</u>. Cambridge, 1929. (287 pp.)
Consists of descriptions of books and quotations from earlier writers.

ORGANIC PRODUCTS

HISTORIES

903. BINZ, Arthur. "Altes und neues über die technischen Verwendung des Harnes," <u>ZAC</u>, 49 (1936), 355-60.
Not much "neues" but an important article.

904. BRACE, Harold W. <u>History of Seed Crushing in Great Britain</u>. London, 1960. (172 pp.)
Much quotation but no citation. The last half is devoted to a list of "known locations of oil mills," and to a list of patents.

905. DICKINSON, H.W. "Charcoal and Pyroligneous Acid Making in Sussex," <u>TNS</u>, 18 (1938), 61-66.
Random information on the subject inspired by the discovery of the site of a former vinegar works.

906. EASTMAN, Whitney. <u>The History of the Linseed Oil Industry in the United States</u>. Minneapolis: Denison, 1978. (272 pp.)
No citations.

907. GILDENMEISTER, Eduard. <u>The Volatile Oils.</u> New York, 1913. (3 vols, trans. from the 2nd German ed.)
Vol. 1 begins with 244 pages of history with profuse citation.

908. HAWLEY, L.F. "Fifty Years of Wood Distillation," <u>I&EC</u>, 18 (1926), 929-30.
Summary without references.

909. LIPPMANN, E.O. von. "Zur Geschichte des Alkohols," in his <u>Beiträge</u>, (no. 130), vol. 2, pp. 5-6, 66-73 from <u>Chemiker Zeitung</u>.
Considers whether alcohol was known prior to the European Middle Ages. He decides in the negative. Citations.

910. MULLIN, B.F., and H.L. HUNTER. "The History of Acetic Acid Manufacture," <u>Cellulose</u>, 1 (1930), 84-89.
Summary without citations.

911. SAVARE, J. "Histoire de l'huile d'olive," <u>Bulletin de la Société Scientifique d'Hygène Alimentaire</u>, 23 (1935) 377-91.
Not seen.

912. SCHMAUDERER, Eberhard. <u>Die geschichtliche Entwicklung der Kenntnisse über die Fette und Öle, Rohstoffquellen, Technologie, Kenntnisse über die Beschaffenheit und Verwendung in der Vorgeschichte und im Alterum.</u> Frankfurt/M, 1964. Dissertation, Univ. of Frankfurt.
Not seen.

PAPER

HISTORIES

913. <u>Papermaking: Art and Craft.</u> Washington: Library of Congress, 1968. (116 pp.)
Catalogue of an exhibit. Summary but authoritative, with numerous illustrations.

914. BOGDAN, Istvan. "Einige technische Daten zur Papiermacherei im 17. bis 19. Jahrhundret." <u>Papiergeschichte</u>, 19 (1969), 36-39.
Compares the time and energy requirement of papermaking over this period.

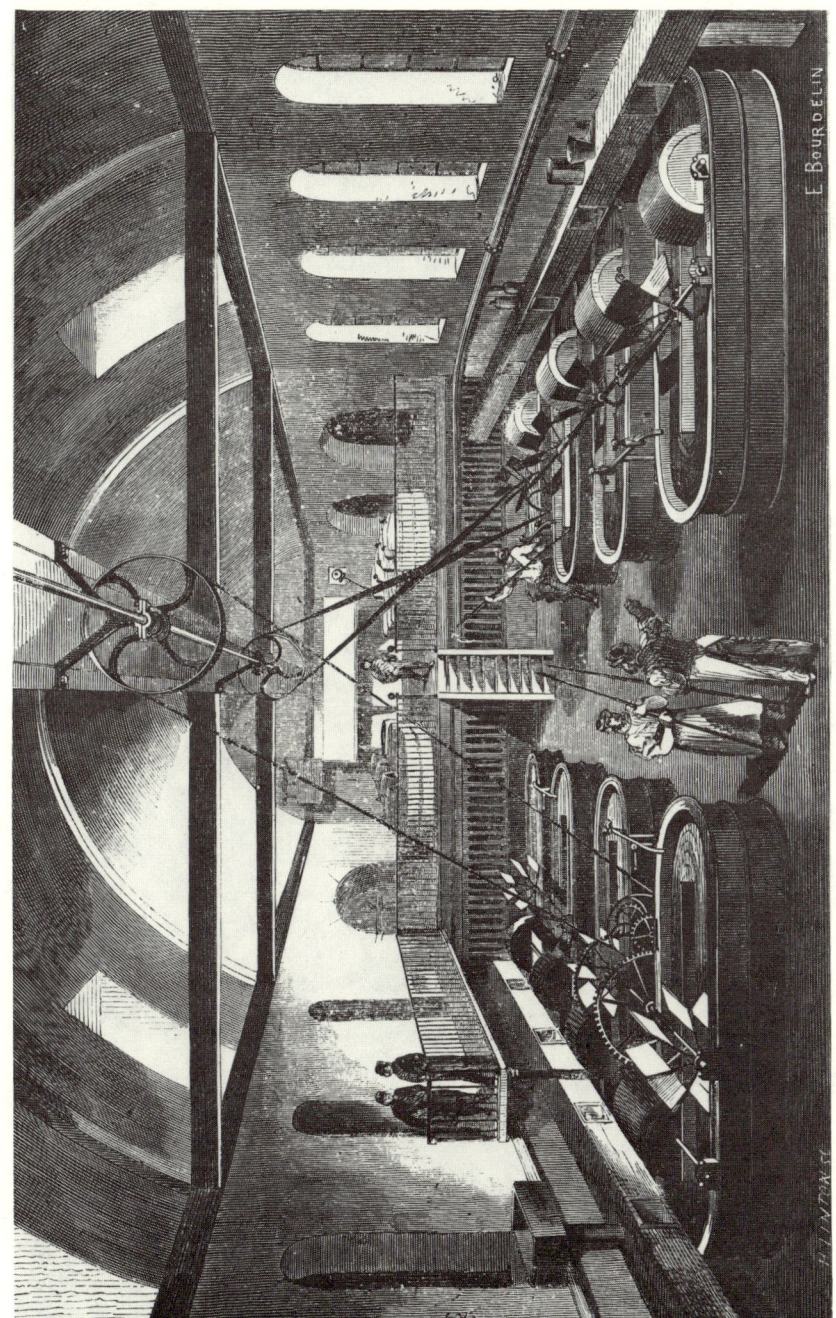

Figure 4. Beaters and washers at the paper factory of Essone, France. From Turgan (item 226)

915. CLASS, Wilhelm. "Von alten papiermühlen." <u>Papiergeschichte</u>, 2 (1952), 16-22, and, as "Wunderwerke alter deutscher Technik," 6 (1956), 1-12.
Describes old papermaking machinery. Citations. Illustrations.

916. DORENFELDT, L.J. "Beitrag zur Geschichte der Sulfatzellstoffabrikation," <u>Der Papier-Fabrikant</u>, 26, (1928), 97-107.
In the last 50 years; a personal account by one who had been involved since 1884. But most of this article is devoted to a reprinting of a pioneer paper, "Vorschriften für die Zellstoffabrikation," (1884) by C.F. Dahl.

917. FELIX, D.A. "What Is the Oldest Dated Paper in Europe?" <u>Papiergeschichte</u>, 2 (1952), 73-5.
Discusses Codex orient. 208 (University of Leyden), dated A.D. 866.

918. GRAAF, Peter. "Zur Entwicklungsgeschichte der Synthesefaser Papier," <u>Papiergeschichte</u>, 17 (1967), 49-61.
Dates the first production of paper from synthetic fibers, 1933, at Windsor Locks, Connecticut. Citations (and samples).

919. HUNTER, Dard. <u>Paper Making through Eighteen Centuries</u>. New York, 1930. (358 pp.)
Narrative without citations but begins with 106 pp. (Pt. I) on the history of papermaking.

920. HUNTER, Dard. <u>Papermaking: The History and Technique of an Ancient Craft</u>. New York, 1943. (444 pp.)
A very nearly definitive work by a long-time student of the subject (who wrote other books on the history of paper in India, China, Japan, Southeast Asia, and early America). Citations. Illustrations.

921. KEIM, K. <u>Geschichtliche Entwicklung der Papierherstellung und der Rohstoffe</u>. Wiesbaden, 1953. (56 pp. Schriftenreihe zur Berufausbildung in der Papierindustrie, no. 2)
An up-to-date summary with references. Notable for its discussion of the variety of raw materials used.

922. LEIF, Irving P. An International Sourcebook of Paper
History. Hamden, Conn.: Archon, 1978 (160 pp.)
This is a bibliography, recent and comprehensive, but
without citations. Includes bibliographies of museums
and of historians of paper.

923. MILLER NICOLASSEN, N.A. "Tycho Brahe und seine Papiermühlen," Papiergeschichte, 4 (1954), 57-67.
An historical-archeological study condensed from the
author's Tycho Brahes papirmille paa Hven (Copenhagen,
1954).

924. MUNSELL, Joel. Chronology of the Origin and Progress
of Paper and Paper-making. Albany, N.Y., 1876.
(263 pp., reprinted by Garland, N.Y., 1980)
One of the most readable but perhaps not unimpeach-
able of chronologies. He informs us, for example,
that "a German by the name of Ebart" invented in 1850
an incombustible and impermeable paper "suitable for
roofing houses."

925. PAPIERGESCHICHTE. (Darmstadt), 1951 ff.
Published by the Verein der Zellstoff- und Papier-
Chemiker und Ingenieure, this outstanding journal is
international in its concerns and in the languages of
publication. Numerous articles are here entered
separately. The early issues contained annual bibliographies by Dard Hunter, later issues include bibliographies of the writings of particular scholars.

926. RUE, -. "50 Years Progress in the Pulp Industry,"
I&CE, 18 (1926), 917-19.
Summary.

927. SANDERMANN, Wilhelm. "Papier in Altamerika," Papiergeschichte, 20 (1970), 11-24.
A detailed history and analysis with citations.

928. SMITH, David C. History of Papermaking in the United
States. New York: Lockwood, 1970. (693 pp.)
Largely economic and business history, where it is
exhaustive. Numerous references.

929. WEEKS, Lyman H. A History of Paper-manufacture in the
United States, 1690-1916. New York, 1916. (352 pp.)

Old fashioned but replete with detail, extensive quotation and citations.

930. WEIR, Thos. S. "Some Notes on the History of Papermaking in the Near East," Papiergeschichte, 7 (1957), 43-8.
Anecdotal. No citations.

931. VOORN, Henk. "Die Anfänge der Papiermacherei in den Niederländen, bis zum Ende des 16. Jahrhunderts," Papiergeschichte, 5 (1955), 23-8.
The earliest mill was at Gennep, in the early 15th century. Describes several early mills. Citations.

932. VOORN, Henk. "Zur Erfindung des Holländers," Papiergeschichte, 5 (1955), 38-42.
Attributes the invention to J.J. Becker, at Zaandem in 1680. Detailed but no citations.

PYROTECHNICS

HISTORIES

933. BROCK, Alan St. H. A History of Fireworks. London, 1949. (180 pp.)
Mostly modern, but deals very briefly and "somewhat superficially," with fireworks before Biringuccio (1540). The author was a manufacturer of fireworks. Citations.

934. FAEHLER, Eberhard. Feuerwerke des Barok: Studien zum öffentlichen Fest und seiner literarischen Deutung vom 16. bis 18. Jahrhundret. Stuttgart: Metzler, 1974. (248 pp.)
Studies of public festivals in the seventeenth and eighteenth centuries. Little technical history but some citations and a bibliography listing 24 early works on this technology.

935. GUTTMANN, Oscar. Monumenta pulveris pyrii. London, 1906. (34 pp. plus 101 plates)
The text is a mere introduction to the plates, but these are the selection of a leading authority and deal more with saltpeter and gunpowder than with firearms. Some indication of sources.

936. HALDEN, J. "A Possible Solution to the Problem of Greek Fire," Byzantinische Zeitschrift, 70 (1977), 90-99.
Suggests that it consisted only of crude petroleum, heated and projected under air pressure.

937. KOTOSKI, Alfons. "Zur Geschichte der chemischen Kampstoffe," ZAC, 51 (1938), 212-14.
A cursory account of chemical warfare as it appears in the works of J.R. Glauber (17 C).

938. LOTZ, Arthur. Das Feuerwerk, seine Geschichte und Bibliographie. Der Feste und der Theatrewesens in sieben Jahrhunderten. Leipzig, 1941. (135 pp.)
Calls a festival at Vicenza in 1379 (described in L.A. Muratori, Rerum Italicarum Scriptores, Milan 1723-51, vol. 13) the first public fireworks display in Europe. Deals with several European countries in turn. Large bibliography, mostly to literature on festival occasions.

939. PARTINGTON, J.R. A History of Greek Fire and Gunpowder. Cambridge, 1960. (381 pp.)
Deals with gunpowder and incendiaries in warfare, in China, Islam, and the Christian west before about 1600. Better described as a mine of detailed and documented information than as a narrative.

940. PARTINGTON, J.R. "The Early History of Phosphorus," Science Progess, 30 (1936), 402-12.
Concerned with the discovery of phosphorus, which he dates 1674-76 and attributes to Hennig Brand of Hamburg. Cites three early accounts, Wilhelm Homberg's in AdS Paris, 10 (1730) Mém. 57-61 (written 1692), Robert Boyle's in PT, 17 (1693) 583 (and in his Aerial noctiluca), and G.W. Leibniz's in Miscellanea Berolinensis, 2 (1710) 91.

941. REINAUD, J.T., and E. FAVE. Du feu Grégeois: des feux de guerre et des origines de la poudre à canon. Paris, 1845. (285 pp.)
A standard work, now outdated. References.

942. ZENGHELIS, C. "Le feu grégeois et des armes à feu des Byzantines," Byzantion, 7 (1932), 265-86.

Greek fire first mentioned in 673. He claims that it contained saltpeter, on the assumption that that substance was mentioned by Pliny and other ancients—an idea not widely accepted. Useful for its citations.

RUBBER AND NATURAL PLASTICS

EARLY WORKS

943. CONDAMINE, C.M. de la. "Mémoire sur une résine élastique, nouvellement découverte à Cayenne par M. Frasneau: et sur l'usage de divers sucs laiteux d'arbres de la Guiane ou France équinoctiale," <u>AdS Paris</u>, (1751) Mém. 319-33.
Celebrated as the first significant description of rubber, a sample of which the author reports having sent to the Academy in 1736.

944. GOODYEAR, Charles. <u>Gum-elastic, and Its Varieties, with a Detailed Account of its Application and Uses, and of Discovery of Vulcanization.</u> New Haven, Conn.: privately printed. 2 vols. 1855.
The author, most successful of those who attempted to improve the utility of rubber through chemical treatment, apologizes for his literary deficiencies and for "alluding to himself oftener than he could wish, by speaking of the trials and discouragements which he had to encounter during the first seven years of his experiments." He can be forgiven, for the book is informative and entertaining. This rare book was reprinted within the pages of the <u>India-Rubber Journal</u> (in a multitude of fragments) in 1936.

HISTORIES

945. "Annals of Rubber," <u>India Rubber World</u>, 90 (1934, July), 41 ff. to 93 (1936, Feb.), 28.
A chronology of rubber, about 21 pages distributed through twenty issues of this journal over three years.

946. "Caoutchouc and Gutta-percha," <u>SIAR</u>, (1865), 206-20.
A chatty and popular but historically informative account of the introduction and early application of these natural plastic materials.

947. "Gutta taban oder Guttapercha," NN, 13 (1860), 222-78.
Dates the first description 1847. Describes its properties and the literature to date. Compares it with rubber, which is also discussed at length, including Goodyear's and other vulcanization processes.

948. BARKER, P.W. "I Have Seen Goodyear's Gum Elastic." Rubber Age, 39 (1936), 155-56, 217-18.
Random comments on Goodyear's rare book (no. 944), from copies in the Library of Congress and the Smithsonian Institution. He says that the latter, bound in "hard rubber," has weathered into a block of rubber (it is no longer at the Smithsonian). In a subsequent issue of this journal (40 [1937] 96-96) the author prints Goodyear's patents.

949. FORBIN, Victor. Le caoutchouc dans le monde. Paris, 1943. (286 pp.)
A comprehensive and readable popular history of natural rubber, mostly economic, social and political. No citations, bibliography, or index.

950. GEER, Wm. C. The Reign of Rubber. New York, 1922 (344 pp.)
A well-written and apparently authoritative history but devoid of documentation. Includes details on manufacture of bicycle tires, balloons and other oddities.

951. GEER, W.C., and C.W. BRADFORD. "The History of Organic Accelerators in the Rubber Industry," I&CE, 17 (1925), 393-96.
From the use of stearic and oleic acids by C.O. Weber in 1904. Summary. No citations.

952. GIBBONS, W.A. "The Rubber Industry, 1839-1939," I&EC, 31 (1939), 1193-1202. (reprinted in SIAR, 1940: 193-214)
A useful summary with valuable citations.

953. HOWARD, Frank A. Buna Rubber, The Birth of an Industry. New York, 1947. (307 pp.)
Primarily an account of the synthetic rubber program of the United States in the Second World War. Few citations but numerous documents are identified and quoted in the text and in appendices. The author was a participant in the program.

954. JUENGER, Wolfgang. <u>Kautschuk. Vom Gummibaum zur Retort</u>. Munich-Vienna, 1952. (201 pp.)
Comprehensive, popular treatment with a brief chapter on synthetic rubber. Useful statistical tables from 1822. Bibliography but no citations.

955. SCHRIDOWITZ, P., and T.R. DAWSON, eds. <u>History of the Rubber Industry</u>. Cambridge, 1952. (406 pp.)
Good analysis (by 35 "contributors"!) of the technologies involved. No citations but a chronology and an annotated bibliography which is very useful.

956. SCHURER, H. "The Macintosh: The Paternity of an Invention," <u>TNS</u>, vol. 28, 1951-53, pp. 77-87.
Charles Macintosh (1766-1843) introduced (in 1823) "the first mass-produced article...into which manufactured rubber entered as an essential ingredient."

957. STERN, H.J. "William Brockedon and the Discovery of Vulcanization," <u>India Rubber Journal</u>, 108 (1945), 615-19, 645-48.
Brockedon (1787-1854) was a partner in the firm of Macintosh from 1845. Author says that he coined the term "vulcanization." No citations.

958. WILSON, Charles M. <u>Trees and Test Tubes: The Story of Rubber</u>. New York, 1943. (352 pp.)
Popular, a wartime book preoccupied with programs for rubber conservation in the United States and proposals for the manufacture of synthetic rubber; more valuable as a source than a history. Substantial bibliography on contemporary technology.

959. WOODRUFF, Wm. <u>The Rise of the British Rubber Industry during the 19th Century</u>. Liverpool: University Press, 1958. (246 pp.)
Primarily economic history but describing the technology. Uses the papers of the family of Stephen Moulton (1794-1880), a temporary immigrant to the U.S. who represented Goodyear in 1842 in exploiting his patent to England. Moulton subsequently founded his own company. The author sees the Americans involved in the establishment of the rubber industry in most of Europe. Many citations.

SAL AMMONIAC (ammonium chloride)

A curious volatile salt (it sublimes) found in volcanoes, it was known to Islamic alchemists in the tenth century or earlier, from a particular volcano in central Asia. They subsequently discovered an organic source, and Europe long obtained it from Egypt, where it was sublimed from the soot of furnaces burning camel dung. It was used to cleanse metal surfaces in fine metal work but also as a source of ammonia--the principal source to the end of the eighteenth century.

EARLY WORKS

960. HOMBERG, Guillaume. "Essais de chimie," AdS Paris, (1702), Mém. 33-52.
Demonstrates by analysis that sal ammoniac is a product of the interaction of spirit of salt (hydrochloric acid) and volatile alkali (ammonia).

961. GEOFFROY, C.J. "Observations sur la nature et la composition du sel ammoniac." AdS Paris (1720), Mém. 189-91. Continued in 1723, Mém. 210-22.
Reveals, on the basis of two communications from Egypt, the source of sal ammoniac in soot.

962. LYELL, C. "Vom Salmiak." AdS Stockholm, 13 (1751), 251-65.
Reviews the history of sal ammoniac and its name (citing a writer who gave 48 names for it). Says the Egyptian process was first made known in a letter of 1716, published in the Nouveaux Mémoires des missions de la compagnie de Jésus in 1718. Reports on European efforts to make it.

963. ALBERTI, W.C. Deutliche und gründliche Anleitung zur Salmiak-Fabrik. Berlin & Leipzig, 1780. (84 pp.)
Surveys the literature and repeats some of the experiments in synthesizing sal ammoniac.

964. WEBER, J.A. Chemische Erfahrungen bey meinen und andern Fabriken in Deutschland. Neuwied, 1793. (224 pp.)
In this and other books this author gives a remarkably circumstantial account of the history of his own several sal ammoniac factories.

965. "Procédé employé par le citoyen Chevremond, à Liège, pour la fabrication du sel ammoniac." Journal des Arts et Manufactures, 1 (1795), 389-95.
Reports Chevremond's process, practiced since about 1781, and part of a French attempt to free itself from dependence on England and Holland for this commodity.

966. SEETZEN, U.J. "Ueber die Gewerbe und Fabriken zu Makkum in Friesland: besonders über die dortige bereitung des Salmiaks, des Glaubersalzes und Englischen Bittersalzes, der gemeinem Kuchensalzes und seinen Hutsalzes." Journal für Fabrik, 21 (1801), 23-35.
An important description, illustrated, of a factory whose owner was "reluctant," but apparently willing, to show it.

967. GEDICKENS, J.C. "Ueber die Entstehung der neuen salmiak und Salz Produktion-Fabrik zu Nussdorf bei Wien," Journal für Fabrik, 22 (1802), 55-66.
In addition to describing the process he names other German manufacturers. This account was noticed and extracted in BSE, 3 (1804), 69-72, and in the Annales des Arts, 19 (1805) 279-86.

968. ROTHE, -. "Ueber Salmiak Fabrikation," Magazin aller neuen Erfindung, 3 (n.d.), 287-91, and 4 (n.d.), 42-49 (both 1806?)
A discussion of economic and other problems with reference to French and English, as well as German factories.

969. ROESLING, C.L. Neue Fabriken-Schule. Pt. 3. Die Fabrik des Salmiaks. Erlangen, 1808. (694 pp.)
This must be the largest book published to that date on an industrial chemical process. It features drawings, in great detail and in color, of the factory proposed. The author was professor at several institutions, including Ulm.

970. COLLET-DESCOSTILS, H.V. "Descriptions de l'art de fabriquer le sel ammoniac," Description de l'Egypt. Etat Moderne, vol. 2, Paris, 1809, pp. 413-26.
Describes the production of sal ammoniac in Egypt, as observed on the Napoleonic expedition in 1798.

971. "Note sur le sel ammoniac que produit une mine de houille incendie," AC, ser. 2, 21 (1822), 158-59.
Reports sal ammoniac to be among the products of a burning coal mine near St. Etienne, France.

HISTORIES

972. MULTHAUF, R.P. "Sal ammoniac: A Case History in Industrialization," T&C, 6 (1965), 569-86.
Describes attempts to synthesize in Europe an imported material then of industrial importance, in part, because it was the major source of ammonia. Citations.

973. RUSKA, Julius. "Sal ammoniac, Nushadir und Salmiak," AdS Heidelberg, Sitzungsberichte, 14 (1923), 23 pp.
An etymological investigation of the names given to sal ammoniac. Citations.

974. RUSKA, Julius. "Die Salmiak in der Geschichte der Alchemie," ZAC, 41 (1938), 1321-24.
Relates references to sal ammoniac in writings by al-Razi (d. c. 923 A.D.) to early accounts of minerals in central Asia. Citations.

975. STAPLETON, H.E. "Sal Ammoniac, a Study of Primitive Chemistry," Memoires of the Asiatic Society of Bengal, 1 (1905), 25-41.
Connects the Arabic discovery of sal ammoniac in organic materials (10th century) with a belief in the efficacy of organic materials in alchemy. Citations.

SALT (sodium chloride)

EARLY WORKS

976. THOELDEN, Johannes. Haligraphia, das ist Beschreibung aller Salz Mineralien. Eisleben, 1603. (316 pp.)
A miscellany, dealing only primarily with sodium chloride, but containing the earliest printed descriptions of a number of saltworks. Thoelden is widely believed to be the real author of the works of the famous "Basil Valentine."

977. JACKSON, William. "Some Inquiries Concerning the Salt Springs, and the Way of the Salt-making, at Nantwich,

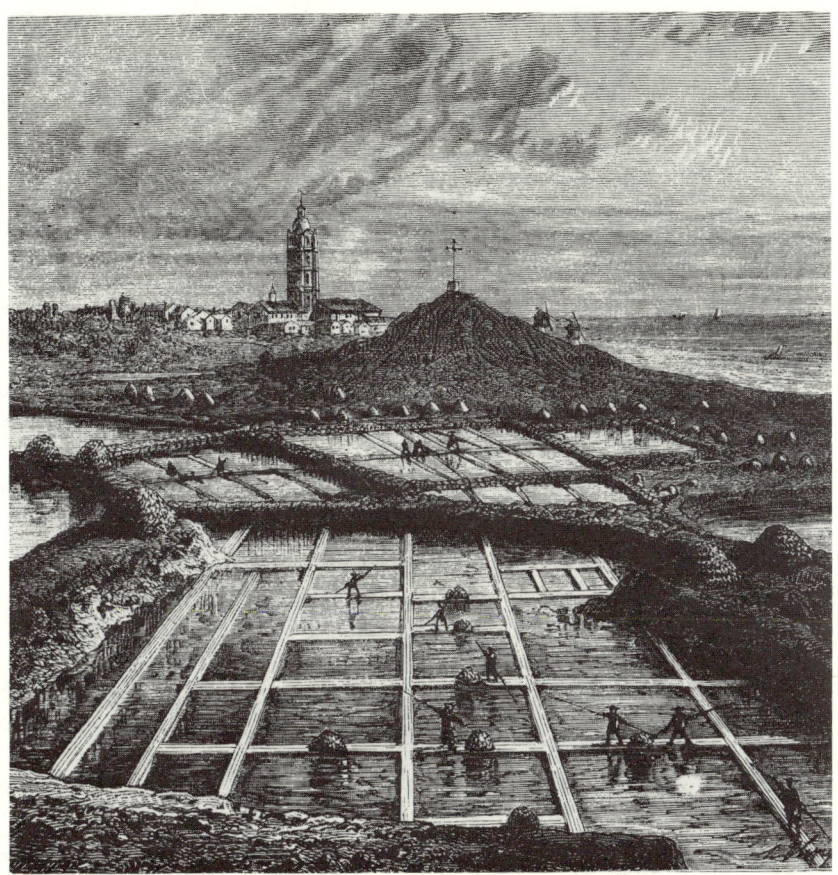

Figure 5. Sea salt production near Croissic, Brittany. From Figuier (item 112)

in Cheshire," PT, 4 (1669) 1060-67.
Response to a questionnaire sent out by the Royal Society of London. One other reply was published, "An Account of the Salt Waters of Droytwich in Worcestershire," by Thomas Rastell (PT, 12 (1679), 1059-64).

978. HONDORFF, Friedrich. Beschreibung des Salzwerks in Halle in Sachsen. Halle, 2nd. ed., 1749. (232 pp. The 1st ed., 1670)

979. KELLNER, David. Berg und Salzwerksbuch. Frankfort & Leipzig, 1701. (562 pp.)
An early compendium of earlier works on salt and other minerals.

980. SWEDENBORG, Emanuel. De sale communi. Philadelphia, 1910. (166 pp.)
Late publication of a manuscript of ca. 1728 by the celebrated Swedish mystic--a detailed description of salt-making which is by no means mystical. Illustrated.

981. BROWNRIGG, William. The Art of Making Common Salt. London, 1730. (295 pp.)
A useful miscellany of British and Foreign information, vaguely aimed to show that (French) "bay salt" could be made in England.

982. DUHAMEL DU MONCEAU, H.L. "Sur la base du sel marine," AdS Paris, (1736), Mém. 215-32.
This article gave the first plausible demonstration that common salt differs from other salts in having "soda" (i.e., sodium) as its "base."

983. WINKELMANN, Heinrich, ed., Das Halleiner Salzwesen. Wethmar/Post Lünen: Gewerkschaft Eisenhütte Westfalia, 1966. (29 pp. and 77 plates)
A useful text by Ernst Penninger and Othmar Schauberger prefaces the reproduction, in color, of 73 oil paintings of salt-making at Hallein, painted about 1757 by Benedikt Werkstätter. An extraordinarily rich source of contemporary information. Privately published.

984. SPENCER, C.F. "Nachrichten von den Salzwerke zu Reichenhall und Traunstein," Beiträge zur Oekonomie, Technologie, Polizey und Kameralwissenschaft. (Göttingen, ed. Johann Beckmann), 8 (1783), 207-42.
A fascinating eye-witness description made by a student of Beckmann who was sent to inspect the saline.

985. DUNDONALD, 9th Earl of (Archibald Cochrane). The Present State of the Manufacture of Salt. London, 1785. (84 pp.)
Aims to demonstrate that salt may be made from rock salt dissolved in sea water at half the cost of making it from sea water alone, that English salt could be equal or superior to foreign salt, and that it could become an important item of export. Discusses the salt tax at length and recommends its abolition. Dundonald is well-known as a (perhaps would-be) pioneer of the British chemical industry.

986. CANCRIN, F.L. von. Entwurf der Salzwerkskunde. Frankfurt, 1788-89. (3 vols)
With this large work (763 pp., with 52 plates) began a series of comprehensive treatises on saltworks published in Germany over the next century. The author (1738-1812/16), son of a mining official, followed the same career himself. His work included the design and operation of saltworks in Germany and finally in Russia. His own son, George, was to hold an important position in the Russian government.

987. RUFFNER, David. "The Kanawha Salt Works," Niles Weekly Register, 8 (1815), 135.
Contemporary description of saltworks near Charleston, West Virginia, by an important producer there.

988. TRAILL, T.S. "On the Salt Mines of Cardona in Spain," Transactions, Geological Society of London, 3 (1816), 404-12.
Describes the mysterious salt mountain in the Pyrenees, a salt quarry which had been to some degree exploited in antiquity but which was still mysterious in the early nineteenth century.

HISTORIES

989. AIGNER, August. "Die Salinen der Alpen in ihrer Geschichtliche Entwicklung," Oesterreichische Zeitschrift für Berg- und Hüttenwesen (Vienna), 36 (1888), 551-57, et seq. (six parts).
Notable for its concern with efficiency. He finds that the "pyrotechnic effect" (fuel efficiency) of the salines in the Salzkammergut improved 33.7% over 379 years. No references.

990. ANCELON, E.A. "Historique de l'exploitation du sel en Lorraine," Mémoires, Académie de Metz, 59 (1879), 153-222.
A circumstantial history of the (mostly medieval) salines of the Seille valley, Moyenvic, Diuze, Vic and Chateau Salins. Citations.

991. ANDREW, E.E. The Early Kanawha Salt Industry, Typescript dissertation, Indiana University, 1936.
Modern circumstantial history with numerous citations.

992. ASHTON, F. "The Salt Industry of Rajputana," Journal of Indian Art and Industry, 9 (1902), 23-32.
Primarily an account of the saline lake Sambhar.

993. BALTZ, Carl. "Die Siedesalz-Erzeugung von ihren Anfängen bis auf ihren gegenwärtigen Stand," ZBHS, 44 (1896), 207-372.
The difficulty of keeping up with the numerous "improvements" described in patents and the literature led the Verein deutschen Salinen und Salzbergwerke to offer in 1881 a prize for an appraisal of improvements made since 1660. The prize was awarded to Baltz, Director at Ischl, whose appraisal is here printed. Important.

994. BATHE, Greville. "The Onendaga Salt Works of New York State," TNS, 25 (1945), 17-26.
Salt production near the present Syracuse, N.Y. began in 1788. A good circumstantial history. Citations.

995. BISCHOF, F. Die Steinsalzwerke bei Stassfurt. 2nd ed. Halle, 1875. (95 pp.)

Deals with the rationale for drilling for salt at Stassfurt, and describes its discovery in 1857 and the consequences. The 1st ed. was in 1864.

996. BLOCH, M.R. "The History of Salt Technology," Actes du VIIe Congrès Internationale d'Histoire des Sciences (Paris), 1953:221-25.

997. BLOCH, M.R. "Salt in Human History," Interdisciplinary Science Review, 1 (1976), 336-52.
The author has published many short surveys of aspects of the subject.

998. BRIDBURY, A.R. England and the Salt Trade in the Later Middle Ages. Oxford, 1955. (198 pp.)
Much useful information, some on technology. Citations.

999. CALVERT, A.F. Salt in Cheshire. London, 1915. (1206 pp.)
The standard history of the most important salt region in Britain. Citations.

1000. CARLE, Walter. "Salsauche und Salzgewinnung im Königreich Württemberg," Veröffentlichungen der Kommission für geschichtliche Landeskunde in Baden-Württemberg (Stuttgart), ser. B, 43 (1968), 105-76.
Summarizes a number of individual articles on the history of salt in the present state of Baden-Württemberg. These and others published subsequently, all done with admirable scholarship, are listed in Multhauf (no. 1032) p. 287.

1001. CRAMER, H. Ein Bruchstuck aus der Geschichte der Königl. Pruess. Saline zu Schönebeck. Schönebeck, 1892. (83 pp.)
Schönebeck, a saline founded in the early eighteenth century, became in the nineteenth century the most productive and technically advanced in Prussia and probably in Germany. This is an interesting circumstantial history.

1002. DE BRISAY, K.W., and K.A. EVANS, eds. Salt: The Study of an Ancient Industry. Report on the Salt Weekend held at the University of Sussex, 22-22 Sept., 1974.

Colchester: Colchester Archaeological Group, 1975. (94 pp.)
25 short reports, chiefly archeological and mostly referring to Britain, but including Niger, France, Belgium, Japan, Colombia and "Africa."

1003. DLUGOSZ, Alfons. Wieliczka. Warsaw, 1958. (152 pp.)
A recent history of the most famous of all salt mines, Wielicza (near Cracow), which is well documented from the early fourteenth century. In Polish, with citations and splendid illustrations.

1004. FICHTEL, J.E. von. Geschichte des Steinsalzes und der Steinsalzgruben im Grossfürstenthum Siebenburgen. Nürnberg, 1780.
Siebenburgen (Transylvania) was the most important source of rock salt in Europe before the nineteenth century but in fact little was produced, and its history, which may go back to 1000 A.D., is very vague. The author reveals that few were familiar with rock salt and that those who were had a wonderful variety of opinions as to its origin. This book is part two of his Beiträge zur Mineralgeschichte von Siebenbürgen.

1005. FLURL, M. "Aeltere Geschichte der saline Reichenhall," Ads Munich, Denkschrift, 1811: 149-90.
This, the oldest Bavarian saline (it may have been Roman), and the most important, was notable for technological "improvement." Salt is still produced there. This history, however, ends with the establishment of a satellite saline at Traunstein (1618).

1006. FOERSTER, Johann C. Beschreibung und Geschichte des Hallischen Salzwerks. Halle, 1799. (262 pp.)
Halle, possibly prehistoric, became the most famous of German saltworks, more for political than for economic or technical reasons. The author acknowledges the earlier publication of Hondorff (no. 978) as a source but goes much further in the detail given on all aspects of the saline.

1007. FREYDANK, Hanns. Die Hallesche Pfännerschaft, 1500-1926. Halle, 1930. (336 pp.)
A thorough and well-documented history, focusing on the Pfännerschaft, the society of owner-operators.

1008. FREYDANK, Hanns. "Die Saline zu Stassfurt," ZBHS, 82 (1934), 207-372.
A good circumstantial history from vague twelfth century references to the discovery of rock salt at Stassfurt in 1851. Citations.

1009. FUERSEN, Otto. Geschichte des kursächsischen Salzwesens bis 1586. Leipzig, 1897. (144 pp.)
This is largely a history of trade and politics, for lower Saxony had no domestic source of salt during this period. The author continued the history in his "Das kursächische Salzwesen seit dem Tod des Kurfürsten August," Neues Archiv für sächsische Geschichte und Altertumkunde (Dresden), 26 (1905), 63-106.

1010. FULDA, Ernst. "Aus der Vorgeschichte des Stassfurter Kalisalzbergbaues," ZBHS, 79 (1931), B232-46, 82 (1934), 155-67.
A detailed account of how and why boring was undertaken at Stassfurt in 1839, and the ultimate consequence, namely, the discovery of mineral potash.

1011. GEHRING, Paul. "Schwäbisch Hall und das Salz," Württembergisch Franken, n.s. 24/25 (1950), 154-79.
A history of the most important saline in Württemberg. Bibliography.

1012. GREVEL, Wilhelm. Die Geschichte der Saline und des Solbades Königsborn. Unna, 1954. (15 pp.)
Established in the 1530's, Königsborn was a typical exemplification of the determined effort made by the German states to achieve self-sufficiency in salt. Like other German salines, it finally survived as a Solbad (health spa).

1013. GUENTHER, Wilhelm. Die Saline Hall i. Tirol, 1272-1967. Vienna, 1972. (97 pp. Leobener Grüne Hefte, no. 132)
A narrative history of the official saline of Tyrol, especially interesting for the attempt made after World War II to ensure its survival through application of the newest technology. It closed in 1967. Bibliography.

1014. HALE, John P. "Manufacture of Salt in Kanahwa," in
G.W. Atkinson, ed., History of Kanahwa County, West
Virginia, Charleston, 1876, pp. 223-49.
Based on the intelligent reminiscences of a senior
resident, this has become the basis for most subsequent histories of salt at Charleston.

1015. HOCQUET, Jean-Claude. Le sel et la fortune de Venise.
Lille: Presses Universitaires de Lille, 1978. (2
vols)
With this massive work (1092 pp.) the role of salt in
the history of Venice (up to 1650) has finally been
given a study commensurate with its importance. Citations, many to archival sources.

1016. HRDINA, J.N. Geschichte der Wieliczkaer Saline.
Vienna, 1842. (274 pp.)
Superceded by Dlugosz (no. 1003) and Kekowa (no.
1019), but still useful for those who cannot read
Polish. Wieliczka was under Austrian control in 1842.

1017. HUGHES, Edward. "The English Monopoly of Salt in the
Years 1563-71," English Historical Review, 40 (1925),
334-50.
On efforts to make England self-sufficient, mostly
political, but including establishment of the ultimately important saltworks on the Tyne. Citations.

1018. HUGHES, Edward. Studies in Administration and Finance,
1558-1825. Manchester, 1934. (528 pp.)
Contains an extended, detailed, and documented history
of the British salt industry.

1019. KECKOWA, Antonina. Zupy Krakowskie w xvi-xviii Wieku
(Salt mines in the Cracow district in the sixteenth -
seventeenth centuries). Warsaw: Akademii Nauk, 1969.
In Polish, with an English summary. Wieliczka and
Bochnia, both dating from the thirteenth century or
earlier, were the principal mines. Gives production
data from 1499.

1020. KLAIBER, Hans. "Die bayerischen Salinen," Saline, 5
(1940), 33-100.
On Reichenhall and its satellites, Traunstein and
Rosenheim. Improves on earlier histories with evidence from an official report of 1764.

1021. KLEIN, Herbert. "Zur älteren Geschichte der Salinen Hallein und Reichenhall," Vierteljahrschrift für Sozial und Wirtschaftsgeschichte, 38 (1952), 306-33.
An important article on Hallein, official saline (and principal source of revenue), of the principality of Salzberg, and on its rivalry with nearby Reichenhall. Citations.

1022. KLEINSCHROD, C.T. Skizze der deutschen Literatur über Halurgie. Munich, 1816. (96 pp.)
Narrative, with historical references; a reminder of the importance attached in former times to salt-making.

1023. KLICKOWSTROEM, C. "Salz, ein bibliographische Studie," Börsenblatt für der deutsche Buchhandel, 16 (1960), 741-44.
A brief but useful bibliography.

1024. KOCH-STERNFELD, J.E. von. Die teutschen, ins besonders die bayerischen und österreichischen Salzwerke, zunachst im Mittelalter. Munich, 1836. (386 pp., reprinted in 1969)
The most comprehensive general history and description of the early saltworks of Germany.

1025. KOERNER, Gerhard. "Das Salzwerk zu Lüneburg," Lüneburger Blätter, 7-8 (1955), 41-55.
One of several articles by this author in this periodical on the most productive saline in medieval Germany. It ceased operation in 1980.

1026. KOPP, U.F. Beytrag zur Geschichte der Salzwerks in den Soden bei Allendorf an der Werra. Marburg, 1788.
The oldest documentary evidence is dated 1491. Soden (or Sooden) was adopted by the state (Hesse) in the sixteenth century as its official saline and subsequently became the center of a small but varied chemical industry.

1027. KURTZ, Heinrich. Die Soleleitung von Reichenhall nach Traunstein, 1617-1619. Munich, 1978. (152 pp. Deutsches Museum, Abhandlungen und Berichte)
A detailed description of technical improvements made at Reichenhall saltworks in the seventeenth century, using manuscript sources in Munich. Citations.

1028. LEMONNIER, Pierre. Les salines de l'ouest. Lille: Presses Universitaires de Lille, 1980. (222 pp.)
A study of the history and present condition of the sea salines on the French Atlantic coast with a view to evaluating technical and social factors in a primitive industry. Citations. An important book.

1029. MARTELL, Paul. "Zur Geschichte des Salzwesens in Westfalen," Kali, 6 (1929), 270-77, 316-24.
Circumstantial historical account of the confusing cluster of saltworks around Unna (including the famous Königsborn). The author had previously published on the salines of various nations, with some historical information, in the same journal, 1910 and 1911.

1030. MERORES, Margarete. "Die venezianisches Salines der älteren Zeit in ihrer wirtschaftlichen und sozialen Bedeutung, "Vierteljahrschrift für Sozial und Wirtschaftsgeschichte, 13 (1916), 71-107.
More on politics and economics than on technology, but an important description of the development of the salt industry which appears to have been a mainstay of the Venetian economy in the years of its ascendency. Citations.

1031. MOLLAT, Michel, ed. Le rôle du sel dans l'histoire. Paris, 1968. (338 pp.)
A series of articles stemming from a conference held in 1961. Very important, especially for ancient and medieval saltworks.

1032. MULTHAUF, R.P. Neptune's Gift: A History of Common Salt. Baltimore: Johns Hopkins Press, 1978. (325 pp.)
The most recent general history. Modesty prevents my praising it. Citations and a bibliography of about 1200 items.

1033. MYERS, J.W. "History of the Gallatin County Salines," Journal, Illinois State Historical Society, 14 (1921), 337-50.
These were the first important trans-Appalachian salines in North America, dating from the beginning of the nineteenth century and probably operated earlier by the "aborigines."

1034. NENQUIN, Jacques. <u>Salt, a Study in Economic Prehistory</u>. Brugges, 1961. (161 pp.)
Important in dealing with prehistoric salt-making, the emergence of identifiable salines, and the economic circumstances involved. Citations.

1035. PARKER, E.W. "History of Salt-Making in the United States," <u>18th Annual Report of the U.S. Geological Survey</u> (Washington) Pt. 5, (1897), 1288-1313.
A sketch, concentrating on production data. Citations.

1036. PENNINGER, Ernst, and G. STADLER. <u>Hallein</u>. Salzburg, 1970 (88 pp.)
Brief circumstantial history of the most productive saltworks in central Europe.

1037. PETER, Charlotte. <u>Die Saline Tirolisch Hall in 17. Jahrhundert</u>. Zurich, 1952. (119 pp.)
One of the most felicitous and detailed accounts of the history of a particular saltworks, largely from unpublished sources. Citations. The saline, near Innsbruck and only recently inoperative, dates from the beginning of the fourteenth century.

1038. POPPE, Gustav. "Zur Geschichte der älteren Saline bei Artern," <u>Zeitschrift des Harz-Verein für Geschichte und Altertumskunde</u>. 1 (1868), 308-17.
Records another attempt to establish self-sufficiency in a German state, unsuccessful, like others, until the application of modern technology.

1039. PRIESNER, Claus. <u>Das deutsche Salinenwesen im frühen 17. Jahrhundert</u>. Munich: Oldenberg, 1980 (66 pp. Deutsches Museum, Abhandlung und Berichte)
A summary based on recent histories of individual salines in Germany. Brief but useful for its citations and illustrations. Gives appropriate attention to technology.

1040. PRINET, Max. <u>L'Industrie du sel en Franche-Comté avant la conquête française</u>. Besançon, 1900. (370 pp.)
Very detailed although unlike most histories it neglects production data. The principal saline was Salins, today Salins-les-Bains. Citations.

1041. NETTLETON, L.C. "History of concepts of Gulf Coast salt-dome formation." <u>Bulletin of the American Association of Petroleum Geologists</u>. 39 (1955), 2373-84.
A rare article on the modern history of salt. It deals with the rock salt sources of Louisiana and Texas, a geological topic originating in the application of chemistry to geology.

1042. NOEL, S.B.J. "Mémoir sur les anciennes salines situées entre les rivières Seine et Arques," <u>Magazin Encyclopédique</u>, 2 (1796), 438-52.
Historical observations on extinct salines in the Dept. of Seine-Inferieure, and the reasons for their disappearance.

1043. PAPY, Louis, "Les marais salants de l'Ouest," <u>Revue Géographique des Pyrénées et du Sud-Ouest</u> (Toulouse), 2 (1931), 121-61.
The French Atlantic coast was the dominant salt producer in Europe for about two centuries after 1450.

1044. RIEHM, Karl. "Prehistoric Salt-boiling," <u>Antiquity</u>, 35 (1961), 181-91.
An analysis of pottery fragments (<u>briquettage</u>) in England, as evidence of early salt-making. The author treated the same topic in relation to central European sites in <u>Jahresschrift für mitteldeutsche Vorgeschichte</u>, 44 (1960), 180-218, and in <u>Forschungen und Fortschritt</u>, 32 (1958), 47-49.

1045. RUF, S. "Zur Geschichte der Saline Hall 1335-61," <u>Archiv für Geschichte und Altertumskunde Tirols</u>, 2 (1865), 184-95.
Deals with an earlier period of the same saline treated by Peter (no. 1037).

1046. SCHLEIDEN, M.J. <u>Das Salz</u>, Leipzig, 1875. (236 pp.)
This informative and charming history by one of the "discoverers" of the animal cell was the product of his resolution to write the "complete" histories of an animal, a vegetable, and a mineral (the horse, the rose, and salt). Only the latter two seem to have materialized.

1047. SCHRAML, Carl. Das Oberösterreichische Salinenwesen. Vienna, 1932-36 (3 vols)
From 1500, the whole being perhaps the most detailed history of the saltworks of Hallein, Hall in Tyrol, Aussee, Ischl, and Hallstatt.

1048. SCHREMMER, Eckart. Technischer Fortschritt an der Schwelle zur Industrialisierung. Munich, 1980. (101 pp.)
An economic analysis of the circumstances surrounding the technical improvement of Reichenhall saltworks in the last decades of the eighteenth century. Citations.

1049. SENF, F.A. "Kurze historisch-technische Beschreibung des Lüneburger Salzwerks," Bulletin des neuesten und wissenswürdigsten aus der Naturwissenschaft (Berlin), 9 (1811), 197-262.
An early historical account by an official of the saline.

1050. SIMMERSBACH, -. "Beitrag zur Geschichte des deutschen Salinenwesens," (Glaser's) Annalen für Gewerb und Bauwesen, Nos. 41-47 (1879), 30 pp.
Important primarily for data on employment and production of the principal German states.

1051. SRBIK, H.R. von. Studien zur Geschichte des österreichischen Salzwesens. Innsbruck, 1914. (229 pp.)
A circumstantial history of the salines of the Salzkammergut, Hallstatt, Hallein, Hall (in Tyrol), Aussee, and Ischl.

1052. VINCENT, Frank. "History of Salt Discovery and Production in Kansas, 1887-1915," Kansas State Historical Society Collections, 14 (1918), 358-78.
Not seen.

1053. VOLGER, W.F. Die Lüneburger Sulze. Lüneburg: Lüneberger Saline, 1956. (112 pp.)
A history written by an official of the saline who died in 1879 but whose book was only recently published.

1054. WEISS, H.B. The Revolutionary Saltworks of the New Jersey Coast. Trenton, N.J.: Past Times Press, 1959. (79 pp.)

An account of attempted sea salines during the American Revolution. The technique was anything but revolutionary.

1055. WEISS, Otto. "Die kurfürstliche Hessische Saline Sooden bei Allendorf," Archiv für Mineralogie, Geognosie, Bergbau und Hüttenkunde (Berlin), 24 (1851), 332-71.
A good history and description, the early history coming primarily from Kopp (no. 1026).

1056. WERNER, C.J. History and Description of the Manufacturing and Mining of Salt in New York State. Huntington, Long Island: by the author, 1917.
Includes a detailed chronology of the early history of salt production at the Onondaga springs (Syracuse).

1057. WITTHOEFT, Harald. "Struktur und Kapazität der Lüneburger Saline seit dem 12. Jahrhundret," Vierteljahrschrift für Sozial- und Wirtschaftsgeschichte, 63 (1976), 1-117.
The most noteworthy attempt known to me to make an accurate determination of the productivity of a saltworks, through economic data and the analysis of ancient weights and measures. Lüneberg was long the most productive inland saline in Europe.

SALTPETER (potassium nitrate) and CHILI SALTPETER (sodium nitrate)

Known in China, and perhaps in India, from the pre-Christian era, in Islam and Europe from the late thirteenth century. Its use in gunpowder overshadowed all other uses until modern times.

EARLY WORKS

1058. THEVENOT, Jean, "Of the Way, Used in the Mogul's Dominions to Make Saltpeter," PT, 1 (1665), 103-04.
Circumstantial description of production at Agra, "extracted" from Thévenot's Relations des divers voyages...Paris, 1663-72.

1059. HENSHAW, Thomas, "The History of the Making of Saltpeter," in Thomas Sprat, The History of the Royal

Society of London, 3rd ed., London, 1722: 260-76 (1st ed., 1667)
One of the first Royal Society essays on the history of trades. Informative, including instructions for producing it.

1060. LISTER, Martin, "A Probable Conjecture about the Origin of the Nitre of Egypt," PT, 15 (1685), 838.
Concludes that the "fossil salt" with which the land abounds may be "made of urinous or salin-urinous" by passing through the bodies of the vast population of crocodiles, hippopotami...(and) "lesser vermine."

1061. HANCOCK, John. Several Methods of Making Salt-petre recommended to the Inhabitants of the United Colonies by their Representatives in Congress. Philadelphia, 1775. (12 pp.)
So that the United Colonies may not need to rely on "foreign importations for gunpowder." Describes a miscellany of methods, some for artificial saltpeter, from European literature. Recommends tobacco barns as the best source.

1062. (FRANCE). "Prix extraordinaire proposé...pour l'année 1778. Programme raisonné de l'Academie des Sciences, pour le prix proposé sur la fabrication du saltpetre & des nitrières." Observations sur la Physique, 6 (1775), 339-46.
The prize, 4000 livres, was authorized by the king in response to a report of the controller of finances on the state of the saltpeter industry. Discusses three theories on saltpeter formation and a Swedish process for "artificial" production, the only one known to the author. This is reprinted in AdS Paris, 1786. hist.:5-12.

1063. ACADEMIE DES SCIENCES (PARIS). COMMISSIONAIRES. Recueil des Memoires et d'Observations sur la Formation et sur la Fabrication du Salpetre. Paris, 1776.
A collection of the writings of twenty-one authors who had written on saltpeter since Glauber. It was translated into German in 1778 by J.F. Pfingston.

1064. WEBER, J.A. Vollständige theoretisch-practische Abhandlung von den Salpeter und den Zeugung

desselben. Tübingen, 1779. (about 234 pp.)
Begins by saying that he knows of no one from Glauber to the present whose writings on saltpeter aren't groundless, a characterisitc comment which, in fact, represents the opinions of other chemists of Weber's own writings. They were numerous; he published a tract of 145 pages on saltpeter in his own short-lived journal (Physikalish-chemisches Magazin, 1780) and a critique in the Paris Recueil (no. 1063) (Anmerkung... über die Sammlunger.., Tübingen, 1780 [120 pp.]). Whatever his demerits, Weber wrote more profusely--and more candidly--than anyone else, on this and other topics in industrial chemistry.

1065. FRANCE. COMITE DE SALUT PUBLIQUE. Programmes des cours revolutionnaires sur la fabrication des saltpêtres, des poudres, et des canons. Paris, an. 2 (1794)
Reports a meeting of L.G. Guyton de Morveau, A.F. de Fourcroy, J.T. Dufourny, C.L. Berthollet, J.A.A. Carny, - Pluvinet. Gaspard Monge, J.H. Hassenfratz, and J.C. Perier, in which each presents his opinions as to the best methods of production.

1066. ANDREOSSY, A.F. "A Report Relative to the Manufacture of Saltpeter and Gunpowder in Egypt," in Memoires Relative to Egypt, London, 1800, pp. 38-43.
Trans. from one of the French publications of the scientific results of the Napoleonic expedition (see no. 594). He says that saltpeter is "worked from veins near Cairo," but also behind "those little emminances of rubbish produced by the avarice and carelessness of the Mamelukes." Describes the method.

1067. PROUST, J.L. "Sur l'utilité du nitrate de soude," Journal de physique, 63 (1806), 59.
An early suggestion of a use (fireworks) for sodium nitrate, known to exist in the Chilian-Peruvian desert since the sixteenth century.

1068. BROWN, Samuel. "A Description of a Cave on Crooked Creek with Remarks and Observations on Nitre and Gunpowder." Trans. American Philosophical Society, 6 (1809), 235-47.

Reports production methods in use at this and other caves in Kentucky, where the earth reportedly contained one pound of saltpeter per bushel.

1069. BOTTEE DE TOULMONT, J.J.A. and J.R.D. RIFFAULT DES HETRES. <u>Traité de l'art de fabriquer la poudre à canon...précédé d'un exposé historique sur l'établissement du service des poudres et saltpêtres en France.</u> Paris, 1811. (537 pp.)
Begins with a detailed history of the Service des Poudres et Salpetres, which they date to about 1540. Treatises on the ingredients of gunpowder follow. A fundamental work.

HISTORIES

1070. <u>Sodium Nitrate Industry of Chili.</u> References to books and magazine articles. Pittsburgh: Carnegie Library, 1908. (11 pp.)

1071. "Zur Geschichte der Chilisalpeters," <u>Jahrbuch der Gesellschaft für Geschichte und Literatur der Landwirtschaft</u> (Bleicherode), 35 (1936) 2-5.
Notes from various sources, mostly on Thaddäus Hanke.

1072. ASLUND, Bengt. "A Rural War Industry: The Impact of Saltpeter Production in Sweden," in International Commission for Military History, <u>Acta No. 2,</u> Washington, 1975: 63-70.
Deals mainly with governmental actions in that country which was widely believed to be most successful in producing saltpeter domestically. The success seems to have been mainly due to legal burdens put on the peasants from the early 16th century.

1073. BALLAND, A. "Les ateliers revolutionnaires du salpetre," <u>Revue Scientifique,</u> ser. 4, 2 (1900), 105-09, 136-43.
A collection of documents from one commune, St. Julien-sur-Reyssouze, Dept. Ain, illustrating the revolutionary program in France in the 1790's.

1074. BERMUDEZ MIRAL, Oscar. <u>Historia del salitre.</u> Santiego, 1963. (456 pp.)
Comprehensive with numerous references.

1075. DONALD, M.B. "History of the Chili Nitrate Industry,"
AS, 1 (1936), 29-47, 193-216.
The best account in English.

1076. FAUST, Burton. Saltpeter Mining in Mammoth Cave,
Kentucky. n.p.: Filson Club, n.d. (1976?), (96 pp.)
A history of one of the more notable "saltpeter caves," which were "scientific curiosities and temporary commercial bonanzas in many countries" in the eighteenth and nineteenth centuries.

1077. GICKLHORN, J. "Thaddäus Haenke's Rolle in der Geschichte des Chili Salpeters," (Sudhoff's) Archiv für Geschichte der Medizin und der Naturwissenschaften. 32 (1940) 337-70.
Analyzes the literary remains of a German physician who went to South America in 1789 and remained as an active scientist and reputedly the first to differentiate clearly sodium nitrate and to attempt to produce it commercially. Includes his description of this and other salts from his Historia natural de Cochabamba (1789).

1078. GIFFEI, Karl A. Die Entwicklung der deutschen Kalisalpeter-industrie unter besonderer Berücksichtigung ihres Standortes. Hamburg, 1926. (diss., 91 pp.).
Not seen.

1079. GREGORY, Karl V. Grundlagen und Entwicklung des Welthandels mit Chilisalpeter, Breslau (diss. 96 pp.)
A summary history, with statistics on production, reference to political aspects of the industry, and comparison to Indian and European saltpeter (both potassium nitrate).

1080. HERNANDEZ CORNEJO, Roberto. El Salitre. Valpariso, 1930. (199 pp.)
An historical "resumé" from the discovery and exploitation of Chili saltpeter. Comprehensive, with quotations and statistics but no references.

1081. LEATHER, J.W., and J.N. MUKERJI. The Indian Salpetre Industry. Calcutta: Supt. Govt. Printing, 1911. (19 pp. Bulletin 24 of the Agricultural Research Institute of Pusa)

A scientific description of the practices of an industry which still existed and which, according to the authors, had never been accurately described.

1082. LENOIR, H. Historique et législation du salpêtre. Les pharmaciens et les ateliers revolutionnaires du salpêtre, 1793-95. Paris, 1922 (234 pp.).
After a brief introduction he gives a history of saltpeter production and legislation during "the terror," with the particular objective of calling attention to the involvement of pharamacists.

1083. MULTHAUF, R.P. "The French Crash Program for Saltpeter Production, 1776-94," T&C, 12 (1971), 163-81.
Deals particularly with the interplay of theory and practice and the attempt to make "artificial" saltpeter.

1084. PAYEN, R. L'évolution d'un monopole: l'industrie des poudres avant la loi du 13 Fructidor, an V. Paris, 1935 (246 pp.)
Not seen.

1085. PRIETO MATTE, J.J. La industria salitrera, su historia... Santiago, 1945. (94 pp.)
Claims that natives of the Chilian-Peruvian desert were exploiting sodium nitrate from 1729. Gives data on the first exports to Europe from 1830.

1086. PRIEUR, A. "Notice sur l'exploitation extraordinaire de salpêtre qui a eu lieu en France, pendant les années 2 & 3 de la république" (i.e., Sept. 1793 to Aug. 1795). AC, 20 (1797), 298-307.
A sketch, but very specific, of the spectacular program which multiplied French saltpeter production severalfold in two years.

1087. RANC, Albert. "La question des nitrates sous la Restauration," Revue Scientifique, 61 (1923), 615-18.
Very brief and undocumented account of French legislation on saltpeter from the sixteenth century.

1088. REISENEGGER, K. Conferencia sobre la historia del salitre. Santiago, 1930.
Not seen.

1089. RICHARD, Camille. La Comité de Salut Publique et les fabrications de guerre sous la terreur. Paris, 1922. (diss. 835 pp. [sic])
Not seen.

1090. SARKER, J.N. "Saltpeter Industry of India in the 17th Century," Indian Historical Quarterly, 14 (1938), 680-91.
Mainly about local monopolies and English and Dutch attempts to circumvent them. The earliest reference to exportation is in 1628.

1091. SCHELER, L. "Lavoisier et la régie des poudres," Revue d'Histoire des Sciences, 26 (1973), 193-222.
Adds a number of Lavoisier's "instructions" on saltpeter production to those already published in Lavoisier's Oeuvres. References.

1092. THIELE, Ottomar. Salpeterwirtschaft und Salpeterpolitik. Tübingen, 1905. (237 pp.)
Essentially a background history for those concerned (at that time) with the problem of the exhaustion of Chili saltpeter "in foreseeable time."

1093. WILLIAMS, A.R. "The Production of Saltpeter in the Middle Ages," Ambix, vol. 22 (1975), 125-33.
Recounts an heroic experiment to test old recipes for artificial production of saltpeter by the weathering of excrements. Concludes optimistically.

1094. ZIEMKE, P.C. "Early Methods of Saltpeter Production," JCE, 29 (1952), 466-67.
Chiefly devoted to production in the Mammoth Cave, Kentucky during the War of 1812.

SOAP & DETERGENTS

HISTORIES

1095. GIBBS, F.W. "The History of the Manufacture of Soap," AS, 4 (1939), 169-90.
A survey from the beginning to recent times.

1096. POUTET, J.J.E. Le traité des savons. Paris, 1828. (100 pp.)

Only incidentally historical. He speaks of the large-scale soap industry of Marseilles, dating it not earlier than the twelfth century, of the various sources of potash and soda, etc.

1097. SCHAEFER, G. "Soap," Ciba Review, no. 56 (1947), 2014-38.
Three historical articles, useful although very brief.

1098. SCHMAUDERER, Eberhard. "Seifenähnliche Produkte im alten Orient," Technikgeschichte, 34 (1967), 300-10. "Seife und seifenähnliche Produkte im klassischen Altertum," Ibid., 35 (1968), 205-22.
Soap was not known as a particular substance in the ancient Orient. Finds the first certain description in the Pseudo-Galen, Liber de simplicibus medicamentis. Citations.

1099. SCHOENE, Manfred. Aus der Geschichte von P3. Düsseldorf, 1970. (143 pp. Schriften des Werksarchiv der Henkel G.M.B.H., no. 2)
"P3" is a designator coined for a series of cleaning materials and their use. It began with a visit to the United States by Hugo Henkel in 1928, to observe the use of sodium phosphate in cleaning, and by 1970 had reached 66 varieties of P3 (which may justify the author's failure to say exactly what it is).

SUGAR

HISTORIES

1100. BAXA, Jacob. Die Zuckererzeugung, 1600-1850. Jena, 1937. (231 pp.)
A sumptuously illustrated history with many citations. Gives particular attention to the Austrian industry.

1101. BAXA, Jacob. Studien zur Geschichte der Zuckerindustrie in Ländern des ehemaligen Oesterreich. Vienna, 1950. (186 pp.)
Primarily economic history. Citations.

1102. BROWNE, C.A. "The Origins of Sugar Manufacture in America." JCE, 10 (1933), 323-30, 421-27.
Circumstantial history with numerous illustrations and no citations.

1103. DEERR, Noel. "Development of the Practice of Evaporation with Special Reference to the Sugar Industry," TNS, 22 (1942), 1-20.
Technical description with illustrations of a series of sugar-boiling pans, mostly from 1827 to 1880.

1104. DEERR, Noel. The History of Sugar. London, vol. 1 1949. (258 pp.)
Organized by areas. Supplements Lippmann (no. 1108) in coming up to date. Data and references. I have been unable to discover additional volumes.

1105. GROTKASS, R.E. Franz Karl Achard's Beziehungen zum Auslände, seine Anhänger und Gegner. Magdeburg, 1930. (68 pp.)
Citations. Reprinted from Centralblatt für die Zuckerindustrie, 11, (1929-30).
Achard (1753-1821) was the pioneer (1799) in beet sugar.

1106. LAFUENTE y POYANES, Mariano. Origin, pregressos y estado actual de la fabricacion del azucar in nuestra costa del Mediterranao. Madrid, 1918.
Not seen.

1107. LEQUIR, -. Histoire de la fabrication du sucre en France. Paris, 1901.
Not seen.

1108. LIPPMANN, E.O. von. Geschichte des Zuckers seit den ältesten Zeiten bis zum Beginn der Rübenzucker Fabrikation. Berlin, 1890. 2nd ed., 1929. (824 pp).
Comprehensive, probably the most complete history of a major chemical industry. Includes a history of prices. Numerous citations. A supplement was published in 1934--"Nachträge und Ergänzungen zur Geschichte des Zuckers," Zeitschrift, Verein Deutscher Zuckerindustrie, 84 (1934), 808-936.

1109. RATEKIN, Mervyn. "The Early Sugar Industry in Hispanola," Hispanic-American Historical Review, 34 (1954), 1-19.
Sugar cane first brought by Columbus in 1493. By 1550 government and society were largely shaped by the sugar industry. Brief treatment of technology. References.

1110. SPETER, Max. "Beiträge zur Wiegengeschichte der Achardischen Rubenzuckerindustrie. Alexander Nicolas Scherers 'Actenstücke und Nachrichten den Runkelrühen-Zucker betreifend," Deutsche Zuckerindustrie, 57 (1932), 28-32.
On Scherer's reports on Achard's factory, printed in Allgemeine Journal der Chemie in 1799. Citations.

1111. STIEDA, -. Franz Karl Achard und die Frühzeit der deutschen Zuckerindustrie. Leipzig, 1928. (218 pp. AdS Saxony, Abhandlung phil.-hist. Klasse 39, no. 3)
Thorough treatment of early beet sugar industry. Citations.

1112. STEARNS, R.P. "The Production of Sugar in Barbados, ca. 1667," AS, 1 (1936), 173-81.
Prints, from the Henry Oldenberg papers in the Royal Society of London, a description received by the Society in response to its inquiries about trades. Anonymous but Stearns thinks it by Ed. Littleton.

PART III

ABRASIVES

HISTORIES

1113. ACHESON, E.G. "Carborundum: Its History, Manufacture, and Uses," JFI, 136 (1893) 194-203, 279-89.
An account by the inventor.

1114. COLLIE, Muriel, F. The Saga of the Abrasive Industry. Greensdale, Mass.: The Grinding Wheel Institute, 1951. (386 pp.)
Informative. The bulk is devoted to the histories of companies in the United States. No citations.

AGRICULTURAL CHEMISTRY

EARLY WORKS

1115. WALLERIUS, J.G. (praes.), G.A. GYLLENBORG (resp.) Agriculturae fundamenta chemica. Uppsala, 1761. (321 pp.)
A dissertation, reportedly the first on this topic. An English summary, published as "A Natural and Chemical Treatise of the Agriculture," appeared in The Rural Library, 1 (1838), 227-66.

1116. TENNANT, Smithson, "On the Different Sorts of Lime used in Agriculture," PT, 89 (1799), 305-14.
Discusses the kinds of lime used and gives a chemical analysis of samples.

1117. DAVY, Humphry. Elements of Agricultural Chemistry. London, 1813. (323 pp.)

1118. RUFFIN, Edmond. "On the Composition of Soils and Their Improvement by Calcerous Manures," American Farmer (Baltimore), 3 (1821), 313-19.
References.

1119. LIEBIG, Justus. Die Chemie in ihrer Anwendung auf Agricultur und Physiologie. Braunschweig, 1840.
A classic work. I have used the English translation, made from the author's manuscript (London, 1842. 409 pp.)

HISTORIES

1120. BLAKEY, Arch F. The Florida Phosphate Industry. Cambridge, Mass.: Wertheim Committee, Harvard Univ., 1973. (197 pp.)
Chiefly economic history, beginning with the discovery of phosphate rock in Florida in 1889 (22 years after its discovery in South Carolina). Citations.

1121. BRAND, Charles J. "Some Fertilizer History Connected with World War I," Agricultural History, 19 (1945), 104-13.
A good circumstantial account of events and persons involved in the American attempt to supply nitrogenous fertilizers during the war. No references.

1122. BROWNE, Chas. A. A Source Book of Agricultural Chemistry. Waltham, Mass.: 1944 (290 pp. Chronica Botanica, vol. 8)
Less a source book than a history, there being a continuous narrative into which quotations are inserted. Useful references.

1123. BURSTYN, Harold L. "Science Pays Off: Sir John Murray and the Christmas Island Phosphate Industry 1886-1914." Social Studies of Science, 5 (1975), 5-34.

Credits the establishment of this profitable industry, in 1899, to the discovery of the mineral by Murray, a naturalist on the Challenger Expedition (1872). Citations.

1124. DUNLAP, Thos. R. DDT. Scientists, Citizens and Public Policy. Princeton: Princeton University Press, 1981. (318 pp.)
Primarily concerned with DDT (dichloro-diphyneltrichloroethane) as a problem in insecticide regulation and as a factor in changes in that regulation.

1125. FRANK, Adolf. "Geschichte der Kalidüngerfabrikation in Stassfurt," ZAC (1893), 325-26.
Corrects an earlier writer in the same journal, stressing the involvement of Liebig, whose letter of Feb. 26, 1865, is here printed.

1126. GILBERT, J.H. Introduction to the Study of the Scientific Principles of Agriculture. London, 1884. (47 pp.)
A lecture, mainly historical, with citations in text. It is somewhat more historical than his earlier "Sketch of the progress of agricultural chemistry." (BAAS Rept. 1880: 1-28).

1127. HALL, A.R. The Book of the Rothamsted Experiments. London, 1905. (294 pp.)
An analysis of the work of J.B. Lawes (1814-1900) and J.H. Gilbert (1817-1901) at Lawes' "home farm," which became the celebrated Rothamsted Agricultural Experiment Station. The principal subjects are soil nitrogen and fertilization. Includes a list of publications from the station, 1843-1905.

1128. HARDING, Thomas C. Two Blades of Grass. A History of Scientific Development in the United States Department of Agriculture. Norman, Okla., 1947. (352 pp.)
A useful history largely devoted to chemistry. A few citations in the text.

1129. KLIPPERT, -. "Die Entwicklung der Technik in der Düngerindustrie," ZAC, 19 (1905), 322-27.
A sketch, dealing principally with superphosphate. No references.

1130. KRISCHE, Paul. "Die Geschichte der Anwendung der Kalisalze in der Landwirtschaft," Kali, 1 (1907), 164-67.

1131. KRISCHE, Paul. "Die Geschichte der Verwertung des Kalis," Kali, 2 (1908), 461-68.
On the use of various potassium salts as fertilizers.

1132. LORIA, Mario. Camillo Cavour e l'industrie chimica die concimi. Turin, 1964. (114 pp.)
Primarily industrial and political history, on the role of Count Cavour (1810-61) in founding in the 1840's the first factory for superphosphate production in Italy. See also Loria's "Cavour and the development of the fertilizer industry in Piedmont," T&C, 8 (1967) 159-77).

1133. McCALLEN, S.E. "History of Fungicides," in D.C. Torgeson, ed., Fungicides. New York, 1967: 1-37.
Calls the period before 1882 "the sulphur era," 1882-1934 "the copper era," since 1934, the era of organic fungicides. The bibliography is large and the text essentially a supplement to it.

1134. McLAUGHLIN, G.A. "History of Pyrethrum," in J.E. Casida, ed., Pyrethrum, New York, 1973: 3-14.
Not seen. The powdered flower heads of some species of pyrethrum (anacyclus pyrethrum) have been used as an insecticide, apparently from about 1876.

1135. MEYER, L. "100 Jahre Agrikulturchemie. Justus von Liebig und das Problem der Aufrechterhaltung der Bodenfruchtbarkeit," Nova Acta Leopoldina (Halle), n.s. 10 (1941), 1-28.
A retrospective view of Liebig's book on agricultural chemistry (no. 1119) in view of subsequent developments in agriculture. No references.

1136. PACKARD, W.G.T. "Superphosphate, its history and manufacture," Trans., Institute of Chemical Engineers, 15 (1937), 21-44.
Reminiscences of a British manufacturer, with particular reference to mechanization of the process from about 1910. Over half devoted to "recent developments." Illustrated.

1137. PURSELL, Carroll W. "The Farm Chemurgic Council and the United States Department of Agriculture, 1935-1939," Isis, 60 (1969), 307-17.
Deals with chemical solutions, and the politics thereof, to the "farm problem" in the United States during the 1930's. Citations.

1138. ROSSITER, M.W. The Emergence of Agricultural Science: Justus Liebig and the Americans, 1840-1880. New Haven: Yale Univ. Press, 1975. (275 pp.)
An important history whose title sufficiently describes the contents. Numerous citations.

1139. RUSSELL, Sir E. John. A History of Agricultural Science in Great Britain, 1620-1954. London, 1966.
Not seen.

1140. SCHUETT, Hans Werner, "Anfänge der Agrikulturchemie in der erstern Hälfte des 19. Jahrhundrets," Zeitschrift für Agrargeschichte und Agrarsoziologie, 21 (1973), 83-91.
An important summary with annotations representing a bibliography of the subject.

1141. SPETER, Max. "Lampadius' Vorschlag der Apatitmehldüngung," Chemiker Zeitung, 57 (1933), 354.
On the use as fertilizer of "phosphorsauren Kalkerde" in 1833 by W.A. Lampadius. This is a supplement to another article in Superphosphate, (1931) 267-77, which I have not seen.

1142. WEBER, Gustavus A. The Bureau of Chemistry and Soils, It History, Activities and Organization. Baltimore, 1928. (218 pp.)
The history (1-110) includes succinct accounts of specific projects (crop chemistry, textile fibre investigations, etc.). Few citations but large bibliography.

1143. WEHLTE, E. "Die Bedeutung der mineralischen Düngung und die Düngmittelindustrie in den letzen 100 Jahren," Technikgeschichte, 35 (1968), 37-55.
An excellent account in terms of production and yields. Citations.

1144. WILEY, Harvey W. "The Relation of Chemistry to the Progress of Agriculture," in Yearbook, U.S. Department of Agriculture, (1899), 201-58.
An "outline" history of "the debt of agriculture to chemistry" in the nineteenth century. No references.

APPARATUS

HISTORIES

1145. BOSCH, Carl. "Ueber die Entwicklung der Hochdrucktechnik bei dem Aufbau der neuen Ammoniakindustrie," in Les prix Nobel in 1931 (Stockholm, 1933), 18 pp.
A personal account without references. Illustrated.

1146. CHRISTOPHE, R. "L'analyse volumétrique de 1790 à 1860: charactéristiques et importance industrièlle. Evolution des instruments," Revue d'Histoire des Sciences, 24 (1971), 25-44.
Volumetric analysis was not very important to science before 1860-70 but was important to technology from 1824 (F.A.H. Descroizilles). Mentions the industries for acids and alkalis, bleaching-powder, sugar, metallurgy. Citations. An important article.

1147. DELIUS, Friedrich, "Chemische Reaktionen unter hohen Druck," in Les prix Nobel in 1931 (Stockholm, 1933), 37 pp.
No references. Illustrated.

1148. DUBRUNFAUT, A.R. A Complete Treatise on the Art of Distilling. 4th ed., London, 1830. (525 pp.)
According to the preface, "Part 1" deals with history, up to (Eduard) Adam (1801). There is no Part 1, only a 272-page text followed by a 253-page appendix, both of which contain miscellaneous historical material.

1149. DUJARDIN, J. Recherches rétrospectives sur l'art de la distillation. Paris, 1900. (236 pp.)
Mainly organized biographically, with an extensive discussion of printed books to 1850.

1150. FELDHAUS, F.M. "Zur Geschichte des Gasometers," Geschichtsblätter für Technik, 5 (1918), 85-86.

Figure 6. Cane sugar factory, Cuba. From Figuier (item 112)

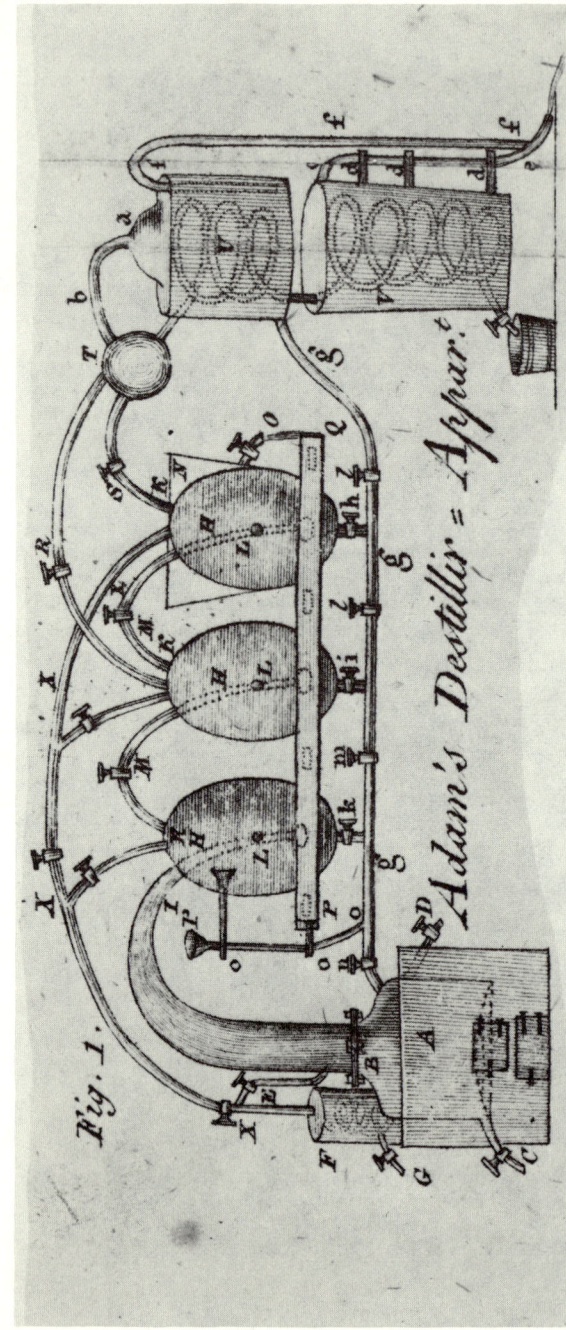

Figure 7. Eduard Adams apparatus for the distillation of alcohol, 1801. From *Dingler's Polytechnisches Journal*, Vol. 2, 1820.

Claims for Christian Huygens (d. 1687) invention of the apparatus credited to Lavoisier a century later, on the basis of an illustration published in the AdS Paris (<u>Machines...approuvées</u>, Paris, 1735, vol. 1, no. 18).

1151. FORBES, R.J. <u>Short History of the Art of Distillation</u>. Leyden, 1948. (405 pp.)
To the mid-nineteenth century. Detailed and authoritative. There is a large bibliography but no footnotes, "which are rightly disliked by the general reader." Many illustrations.

1152. GOODFIELD, J., and S. TOULMIN, "The Quattara: A Primitive Distillation and Extraction Apparatus Still in Use," <u>Isis</u>, 55 (1964), 339-42.
Describes and illustrates a still found in use in Fez, Morocco, in 1963, of a type used in Islam in the Middle Ages. Citations.

1153. LEVEY, Martin. "The Earliest Stages in the Evolution of the Still," <u>Isis</u>, 51 (1960), 31-34.
A vessel excavated at Tepe Gawra (4th millenium B.C.), now in the Univ. of Pennsylvania Museum, is identified as a still.

1154. LIPPMANN, E.O. von. "Geschichte der Kaltemischungen," <u>ZAC</u> (1898), 739-45.
A survey of literary references to cooling mixtures to the end of the eighteenth century. Numerous citations.

1155. RAICHLE, L. "Die Entwicklung der Hockdrucktechnik in der BASF," <u>Erdöl und Kohle</u>, 13 (1960), 601.
A "press release" but a useful description of equipment used in early work at BASF. No citations.

1156. SPETER, Max. "Gegenstromprinzipkühlschlange um 1775." <u>Chemischen Fabrik</u>, 5 (1932), 284.
Refers to the mention of a counter-current condensor by Demachy in no. 18.

1157. SUDHOFF, Karl. "Weiteres zur Geschichte der Distillationstechnik," <u>AGNT</u>, 5 (1914), 282-88.
Discusses cooling apparatus, of Leonardo da Vinci, Philipp Ulstad, and a newly discovered source, Johannes Wenod (c 1420). Citations.

1158. TAYLOR, F.S. "The Evolution of the Still," <u>AS</u>, 5 (1945), 185-202.
A comparison of stills in the Greek alchemical corpus with Renaissance stills.

1159. TIMM, Bernard. "Die Anwendung hoher Drücke in der chemischen Technik," <u>Die BASF</u>, 17 (1967), 23-33.
An "overview" of the economically important processes, for ammonia, n-butylaldehyde, n-butanol, higher alcohols, acetic acid, propionic acid, hydrogenated benzol, acrylic acid, and methanol. No citations.

1160. UNDERWOOD, A.J.V. "The historical Development of Distilling Plant," <u>Transactions, Institute of Chemical Engineers</u>, 13 (1935), 34-62.
A well-done summary, with citations and illustrations, although superceded by Forbes (no. 1151).

1161. WITSCHAKOWSKI, Walter. <u>Hochdrucktechnik</u>. Ludwigshafen, 1974. (107 pp. 5B no. 12)
From 1910, in the synthesis of ammonia, methanol, fuel hydrogenation and products of acetylene and ethylene.

ARTIFICIAL FIBERS and PLASTICS

HISTORIES

1162. ALLEN, J.A. <u>Studies in Innovation in the Steel and Chemical Industries</u>. Manchester, 1967. (256 pp.)
Deals with polythene (polyethylene, developed at ICI ca. 1933), terylene (dacron, developed by J.R. Whinfield and J.T. Dickson, near Manchester, 1941), and oxygen steelmaking. Citations, mostly to patents.

1163. AREMARIUS, J.C., and J.G. KEPPLER. <u>Plastics veroveren de wereld</u>. Deventer, 1950. (352 pp.)
Of historical value for its chronology of events in the history of plastics, 1781-1949. Otherwise not historical although there is a good index facilitating the chemical identification of commercial plastics.

1164. BARTH von WEHRENALP, Erwin, and H.J. SAECHTLING. <u>Jahrhundret der Kunststoffe</u>. Düsseldorf, 1952. (564 pp.)

Begins with a six-page historical introduction with useful tables and graphs. The rest is not historical, being mostly photographs and captions (in German, French, Spanish and English).

1165. BAUER, Robert. Das Jahrhundret der Chemiefassern. Munich, 1951. (298 pp.)
A popular history without references. Comprehensive but thin in detail. Contains a chronology and some production statistics.

1166. BAUER, Robert. Galilith: die Geschichte eines Kunststoffes, 1904-54. Hamburg, 1954.
Not seen.

1167. BEER, Edwin J. The Beginning of Rayon. Paignton (Devon): privately printed, 1962 (206 pp.)
A chatty and undocumented personal account ("mostly from old letters and diaries") by an associate in the 1890's, of C.F. Cross and E.J. Bevan. Most of it is occupied by a rather chaotic but fascinating diary. Part of it was printed as "The Birth of Viscose Rayon," TNS, 35 (1962-63), 109-16.

1168. BOLTON, E.K. "The Development of Nylon," I&EC, 34 (1942), 53-58.
A circumstantial account by a representative of the developer (Du Pont). Dates the research from 1929. No references.

1169. BRANDENBURGER, Kurt. Im Zeitalter der Kunststoffe. Munich, 1938. (100 pp.)
Minimal history but good on the state of the art at the time.

1170. BUCK, A.J. "The Historical Development of Phenolformaldehyde Resinous Products," British Plastics, 7 (1936), 499-500; 8 (1937), 54-6 et seq.
Informative although without citations. Divided into no less than twenty parts, each of 2-3 pages, spread through issues of 1936 and 1937.

1171. DREFUS, H. "The Birth, Development, and Present Position of the Cellulose-acetate Artificial Silk Industry in This Country," Journal, Society of Dyers and

Colourists, 55 (1939), 116-24.
An address, without references, "this country" being Britain.

1172. DUBOIS, J. Harry. Plastics History: USA. Boston, 1972. (447 pp.)
Popular with many illustrations.

1173. DU PONT, E.I. de NEMOURS, firm. Nylon--The First 25 Years. Wilmington: Du Point, 1963.
Not seen.

1174. FAUQUET, L.G. Histoire de la rayone et des textiles synthétiques, Paris, 1960. (269 pp.)
Not seen.

1175. FISCHER, Hugo. "Frederick Walton und die Linoleumindustrie," Technikgeschichte, 19 (1929), 133-36.
Walton (1834-1928), of Yorkshire, received the first of several patents in 1857 and began in 1864 to manufacture the material he named "linoleum." It consisted of a composition of linseed oil (German Leinöl), resin, and Kauriharz (ground cork?), impregnated onto jute or another fabric.

1176. FREEMAN, C. "The Plastics Industry: A Comparative Study of Research and Innovation," Economic Review, National Institute of Economic and Social Research, 26 (1963), 22-62.
An excellent study of the industry with a strong historical bias. Citations.

1177. FRIEDEL, Robert. "Parksine and celluloid, the failure and success of the first modern plastic," History of Technology, 4 (1979), 45-62.
Attributes the failure of the former and the success of the latter to the Hyatts' appreciation of the technical limitations and commercial possibilities of the material. Citations.

1178. FRIEDEL, Robert. Imitation and Artifice: A History of Celluloid. Madison: Univ. of Wisconsin Press, 1983.

The technical history given here is a prelude to a longer discussion of the economic and social history, which the author sees as the key to "the first artificial plastic." Citations.

1179. GARDNER, W.H. "History of the Use of Shellac as a Plastic," British Plastics, 6 (1935), 459-60.
Brief but informative. Citations.

1180. GIBSON, Reginald O. The Discovery of Polythene. London, 1964. (30 pp. Royal Institute of Chemistry lecture series)
A short but perceptive description of the research and development work that led to the exploitation of polythene (polyethylene), written by one of the participating chemists.

1181. HARD, Arnold H. The Romance of Rayon. Manchester, 1933. (76 pp.)
A summary without citations and mainly restricted to Britain. "Compiled" from information supplied by the families of the founders of the industry.

1182. HOECHST, firm. Zur Strukturaufklärung der Makromoleküle. Ein Briefwechsel zwischen Prof. (Hermann) Staudinger und Dr. Kränzlein. Frankfurt/M, 1966. (77 pp., DHA No. 15)
Chiefly from the period 1926-28. Staudinger (1881-1965) won the Nobel Prize in Chemistry (1953) for his work on "giant molecules."

1183. HOECHST, firm. Die Anfänge der Kunstwerkstätte in Hoescht. Frankfurt/M, 1969. (123 pp., DHA no. 40)
Documents, 1928-34, chiefly concerned with the development of corrosion-resistant piping.

1184. HOELSCHER, Friedrich. Kautschuke. Kunststoffe. Fasern. Sechs Jahrzehnte technische Herstellung synthetischer Polymerer. Ludwigshafen, 1972. (145 pp., SB no. 10)
On the history of these materials, especially at BASF. Chronological table.

1185. HOETTENROTH, Valentin. Artificial Silk. London, 1926. (421 pp.; trans. from German ed. of 1926)

A standard technical work, which begins with 18 pp. of history. No citations but a good bibliography.

1186. HOLLANDER, Samuel. "The Sources of Increased Efficiency: a Study of Du Pont Rayon Plants," T&C, 7 (1966), 121-23.
Considers "technical change," in which he includes changes in organization, "overwhelmingly important" in maintaining an increase in productivity of 4-5% per annum over "8 to 24" years. Citations.

1187. HOPFF, H. "Zur Entwicklung der Kunststoffchemie," Chemische Industrie, 4 (1952), 715-25.
Not seen.

1188. KAUFMAN, M. The First Century of Plastics: Celluloid and Its Sequel. London: The Plastics Institute, n.d. (1962). (130 pp.)
A far more important book than one would expect in a centenary volume sponsored by a trade association. The first half deals with celluloid, and the author considers Alexander Parkes to have been the "founder" of the industry, against the often-preferred claims of J.W. Hyatt. The second part deals with other plastics, to the present, being based on the author's "distillation of a number of specialist accounts." Informative, readable, documented, and illustrated.

1189. MARX, Carl, et al. "With the Pioneers." Plastics, 2 (1926), 9-10 et seq.
A series of summary articles without references, namely Carl Marx, "Schoenbein, Discoverer of Cellulose Nitrate," (2 [1926] 9-10, 30-32); Julia Greenfield, "Alexander Parkes and Parkesene" (Ibid., 49-50, 60); Carl Marx, The Case of (Daniel) Spill vs. the Celluloid Company," (Ibid., 86, 97, 100); Carl Marx, "Blood and Shellac Plastics 50 Years Ago," (3 [1927] 462); Carl Marx, "The Rise of the First Great Plastics Industry (i.e., The Celluloid Co.)," (4 [1928] 669-71, 684, 693-94).

1190. MIENES, Karl. Vom Celluloid bis zum Polyaddukt. Munich, 1953. (82 pp.)
Although useful as a guide (with citations) to the realm of plastics at this time, this is no history.

Perhaps the title is intended to refer to the variety rather than the chronology of plastics.

1191. MUTHESIUS, V. Zur Geschichte der Kunstfaster. Heppenheim, 1949. (167 pp.)
Not seen.

1192. ORCHIN, Milton. "A Case History of Transmuting an Idea into Money," JCE, 55 (1978), 782.
On the discovery in 1960 of trimethyldiamine, the catalyst for the preparations of polyurethanes.

1193. SAECHTLING, Hansjürgen. Werkstoff aus Menschenhand: Technik und Wirtschaftsgeschicte, 1910-1960. Munich, 1961. (70 pp.)
This "slim" book is loaded with information, tables, and graphs on the recent history of plastics. Citations.

1194. SCHWEN, Gustav. "Ein halbes Jahrhundert Grenzflächenchemie," Chemische Industrie, 20 (1968), 33-37.
A retrospective account on the sixtieth anniversary of "Nekal" (diisopropylether) as developed at BASF. Describes that and other products developed subsequently.

1195. SPROXTON, F. "The Rise of the Plastics Industry," Chemistry and Industry, 57 (1938), 607-16.
A summary account, from 1827, by the Director of the British Xylonite Co., the first successful British celluloid manufacturer, founded about 1877. Illustrations and statistics but no citations.

1196. STOERI, Fritz. Der Stoff, aus dem die Schäume sind. Die Geschichte vom Styropor. Ludwigshafen, n.d. (1978 ?) (SB no. 16)
Styropor (polystyrol), made by BASF since 1951, is 98% air, and this volume has about the same proportion of advertising--a new format for this series, designed, as the foreword says, to "free it from the smell of the archives."

1197. URQUHART, A.R. "Cellulose Derivitive Rayons, Past, Present, and Future," in The Development of Some Man-made Fibres. Manchester: Textile Institute, 1952.
Not seen.

BLEACHING

    EARLY WORKS

1198. HOME, Francis. <u>Experiments on Bleaching</u>. Edinburgh, 1756. (214 pp.)
With alkali. Includes 37 pages on the various kinds of "ash" and their production.

1199. D.B. "(Letter on) Bleaching Linen Yarn and Linen Cloth," <u>Universal Magazine</u>, 18 (1756), 17-20.
Describes a process for bleaching with ashes.

1200. EASON, Alexander, "Observations on the use of acids in bleaching linen," <u>Memoires, Manchester Literary and Philosophical Society</u>, 1 (1785), 240-42.
Tells us that use of sour milk has long been known: the use of "fossil" (i.e., mineral) acids was introduced about 30 years ago. Nitric acid is corrosive and expensive and sulphuric acid is universally used. Hydrochloric acid is now nearly as cheap, and the author thinks that it will replace sulphuric.

1201. BERTHOLLET, C.L. "Description du blanchiment des toiles par l'acide muriatique oxigéné," <u>AC</u>, 2 (1789), 151-90.
Describes his experiments which led to the introduction of chlorine bleaching commercially; credits the genesis of the idea to Scheele.

1202. RUPP, T.L. "On the Process of Bleaching with the Oxygenated Muriatic Acid," <u>Repertory of Arts and Manufactures</u>, vol. 10 (1789), 165-80. (from the <u>Memoires, Manchester Literary and Philosophical Society</u>).
After a brief historical introduction he gives his own experiments and apparatus. Illustrated.

1203. TENNANT, Charles. "Patent (Jan. 23, 1798) for His Method of Using Calcerous Earth...Instead of Alkaline Substances, for Neutralizing the Muriatic Acid Used in Bleaching," <u>Repertory of Arts and Manufactures</u>, 9 (1798), 303-09.
Prints the specification, followed by a brief elucidation by one of Tennant's associates.

1204. LOYSEL, J.B. "On Bleaching the Pulp for Manufacturing Paper," Repertory of Arts and Manufactures, 16 (1802), 200-15. (from AC 39 [1801])
Principally a description of the apparatus used commercially for Berthollet's process for bleaching rags for paper. Illustrated.

1205. (BERTHOLLET, C.L.) "Note sur les blanchisseries berthólleenes," BSE, 2 (1803), 49-52.
Describes the consequences of Berthollet's work from 1785-86.

1206. LEBLANC, Nicolas. "Observation sur la confection et l'usage de la soude," AC, 50 (1804), 92-106.
Written to counter a prejudice against the use of artificial soda for bleaching linen.

1207. "Ueber die Bereitung eines Salzes zum Bleichen und Waschen," Magazin aller neuen Erfindungen, Entdeckungen, und Verbesserungen für Fabriken, 6 (1806 ?), 245-46.
Reports that a Belgian who had purchased some of Tennant's bleaching powder (Kalch gesättigte Salzsäure) was told, when he reordered, that the king had forbidden its export. He had a few ounces analyzed in Paris and reports that it can now be produced elsewhere.

1208. PAYEN, Anselme. "Procédé pour fabriquer le muriate de chaux," Annales de l'Industrie, 10 (1823), 293-95.
Describes three processes in commercial use, plus a fourth, his own, used in his factory "for some years."

HISTORIES

1209. GREUP-ROLDANUS, S. Geschiedenis der Haarlemmer bleckerijen. The Hague, 1936. (355 pp.)
The history of "Haarlem bleach" from the earliest record (1494). Comprehensive, with many citations. English summary.

1210. HIGGINS, S.H. A History of Bleaching. London, 1924. (176 pp.)
To the mid-nineteenth century and relating primarily to England. Summary but informative with sources only occasionally mentioned in the text.

COAL

EARLY WORKS

1211. LAMBERVILLE, Charles de. <u>Alphabet des terres à bruler et à charbon de forge.</u> Paris, 1638.
Not seen.

1212. MORAND, J.F.C. <u>Art d'exploiter les mines de charbon de terre.</u> Paris, 1769-79 (6 pts.)
Part of the <u>Descriptions</u> (no. 13), this massive discussion of coal is even yet hardly surpassed in size and circumstantial detail. Describes procedures in various regions of France and in England, exploration, mechanization, and the use of coal in metallurgy but does not appear to mention other chemical considerations. Contained in vols. 16-18 of the Neuchatel edition, which adds notes and a chapter on the "prejudices" against coal firing.

1213. PFEIFFER, H.H. von. <u>Geschichte des Steinkohlen und des Torfs.</u> Mannheim, 1775.
Not seen.

1214. DUNDONALD, 9th Earl of (Archibald Cochrane). "Patent (Apr. 30, 1781) for a Method of Extracting Tar, Pitch, Essential Oils, Volatile Alkali, Mineral Acids, Salts, and Cinders from Pit Coal," <u>Repertory of Arts and Manufactures</u>, 1 (1794), 145-48.
Coal burns in a closed vessel with some access to air. It "throws off" the materials mentioned, which are condensed with water cooling. Steam injection completes the process. He later (Ibid. 8 (1798), 79-90) reports a related patent covering the use of coal and soot as manure, rendering coal soluble in water, obtaining volatile alkali from peat, etc.

1215. DUNDONALD, 9th Earl of (Archibald Cochrane). <u>Account of the Qualities and Uses of Coal Tar and Coal Varnish.</u> London, 1785. (43 pp.)
This followed his work of the same title published in Edinburgh, 1784 (23 pp.), which I have not seen. A good circumstantial account of Dundonald's and other British processes appeared in Germany in 1790 (D.L.G. Karsten, "Ueber des Grafen Dundonalds Behandlung des Steinkohlen," <u>Bergmannisches Journal</u>, 1 (1790), 212-25.

HISTORIES

1216. BRIGGS, H. "History of the Coal Oil Industry," <u>Trans., Institute of Mining Engineers</u> (London), 81 (1930-31), 489-506. 82 (1931-32), 25-35.
Begins with J.J. Becher's alleged production of oil by the low temperature distillation of coal (1681) but is chiefly concerned with Britain to the end of the nineteenth century.

1217. FEILER, Paul. <u>Die Wirbelschicht. Einer neuer Aggregatzustand.</u> Ludwigshafen, c. 1972 (83 pp., SB no. 9) On the development of the fluidization of solids by Fritz Winkler at BASF in 1921.

1218. HUGHES, T.P. "Technical momentum in history: hydrogenation in Germany 1898-1933," <u>Past and Present</u>, no. 44 (1969), 106-32.
Citations. An important article.

1219. SCHNEIDER, Eugene. <u>Le charbon. Son histoire, son destin.</u> Paris, 1945. (350 pp.)
On its economic and technical development, including the pre-modern period. Pp. 305-29 deal with the recent history of coal in the chemical industry. Citations.

1220. WEISS, Alfred. "Zur Geschichte der Veredelung und Verwendung steirischer Braunkohlen. <u>Blätter für Technikgeschichte</u> (Vienna), 39/40 (1977/78), 27-46.
Coal was discovered in Styria in 1606, "recification" and coking pursued in the early eighteenth century, alum and sulphur recovery by the late eighteenth century, gasification tried about 1840. Comes up to date. Citations.

DYES, SYNTHETIC

HISTORICAL

1221. BAEYER, Adolf von. "Zur Geschichte der Indigo-Synthese," <u>Berichte, DCG</u>, 33, 3 (1900), Anlage V: 1-lxxx.
An account of the epochal synthesis by the chemist who accomplished it. Citations.

1222. BEER, John J. "Coal Tar Dye Manufacture and the Origins of the Industrial Research Laboratory," Isis, 49 (1958), 123-31.
Finds the prototype of the modern industrial research laboratory in the German dye-works laboratories just prior to the first World War.

1223. BEER, John J. The Emergence of the German Dye Industry. Urbana, Ill., 1959. (168 pp.)
Authoritative, scholarly, and readable, this dissertation on a crucial phase in the history of industrial chemistry exemplifies what is most needed in this field. Citations.

1224. BOLLEY, P.A. Altes und neues aus Farbenchemie und Färberei. Ueberblick der Geschichte und Rolle der sogenannten Anilinfarben. Berlin, 1867. (35 pp.)
Not seen.

1225. BRIGHTMAN, R. "Perkin and the Dyestuff Industry of Britain," Nature, 177 (1965), 815-21.
Discusses Perkin's work in relation to contemporary academic and other institutions.

1226. BRUNCK, Heinrich. "Die Entwicklungsgeschichte der Indigo-Fabrikation," Berichte DCG, 33, 3 (1900) Anlage V, lxxxi-lxxxvi.
Supplements no. 1221. This appears to have been translated as "The History of the Development of the Manufacture of Synthetic Indigo" (N.Y., 1901), but I have been unable to locate it. It is summarized in Chemical News, 86 (1902), 211-13.

1227. CARO, Heinrich. "Ueber die Entwicklungsgeschichte der Teerfarben-Industrie," Berichte DCG, 25, 3 (1892), 955-1105 (also published separately, Berlin, 1895. 115 pp.)
An important account by a pioneer in the industry. Citations.

1228. CLIFFE, W.H. "In the Footsteps of Perkin," Journal of the Society of Dyers and Colourists, 72 (1956), 363-65.
An interesting account of an attempt to make mauve, following Perkin's patent of Aug. 2, 1856. Citations.

1229. CLIFFE, W.H. "A Historical Approach to the Dyestuffs Industry," Journal of the Society of Dyers and Colourists, 79, (1963) 353-63.
Concludes that court decisions of 1876 against a dye tariff accounted for the decline of the British synthetic dye industry. Much detail but no citations.

1230. CROSSLEY, M.L., et al. "Progress in the Dye Industry in the United States during the Past Decade," I&EC, 18 (1926), 1322-46.
A series of short papers read at the annual meeting of the American Chemical Society. K.H. Klipstein, "Development of Synthetic Anthraquinone," (1327-29), is the most important.

1231. DUISBURG, Carl. "Die Wissenschaft und Technik in der chemischen Industrie, mit besonderer Rücksichtigung der Teerfarbenindustrie," ZAC, 25 (1912), 3-14.
Reminiscences of a leader. No references.

1232. FIESER, Louis F. "The Discovery of Synthetic Alizarine," JCE, 7 (1930) 2609-33.
A detailed account of the discovery. Citations.

1233. FORDEMWALT, Fred. "The History of the Development of Fast Dyes," in White, 1956, (no. 1262), 349-56.
A too-brief summary but containing formulas for early dyes and tables of the number of dyes "discovered" by decades, from 1860 to 1910. No references.

1234. GARDNER, W.M., ed. The British Coal-tar Industry: Its Origin, Development, and Decline. London, 1915. (437 pp.)
A source book, reprinting, or extracting (or sometimes simply mentioning) 32 articles published between 1868 and 1915. Some are specifically historical (and are separately mentioned here) and all are of historical interest.

1235. HOECHST, firm. Die entscheidenden Jahre der Indigo-Synthese. Frankfurt/M, 1967. (100 pp. DHA no. ?)
Not seen.

1236. HOECHST, firm. Beginn der Alizarine-Aera. Frankfurt/M, 1970 (66 pp., DHA no. 42)

Consists of documents and reprinted material from 1868, relating to the work of Carl Graebe and C.T. Liebermann.

1237. HOLZZCH, K. Entwicklungsarbeit der deutschen chemischen Industrie auf dem Gebiet der Farbstoff-Synthese. Stuttgart: IG Farbenindustrie, 1936.
Not seen.

1238. LAUTERBACH, Fritz. Geschichte der in Deutschland bei der Färberei angewandten Farbstoffe. Leipzig, 1905. (113 pp.)
Not seen.

1239. LESCHER, H.Z. "Early Synthetic Dyes," in White, 1956 (no 1262), 297-305.
Lots of information in telegraphic style with much citation.

1240. LEVENSTEIN, Ivan. "The Development and Present State of the Alizarine Industry," Journal, Society of Chemical Industry (London), 2 (1883), 213-27.
A brief historical sketch is followed by a detailed account of production methods. No citations.

1241. LIGHTFOOT, J. The Chemical History and Progress of Aniline Black. Burnley, 1871.
Not seen.

1242. MEDOLA, R. The Founding of the Coal-Tar Colour Industry," Journal, Society of Dyers and Colourists, 24 (1908), 95-105. (reprinted in Gardner, no. 1234: 234-56)
Relates particularly to the work of Perkin, who had died the previous year. No citations.

1243. MORTON, James, "History of the Development of Fast Dyeing and Dyes," Journal, Society of Arts (London), 77 (1929), 544-74.
On the development of fast dyes in the early twentieth century. The author, who learned dye chemistry at Leipzig in 1869, was personally involved.

1244. NAGEL, Alfred von. Fuchsin. Alizarin. Indigo. Frankfurt/M, 1970 (59 pp. SB no. 1)
Histories of dyestuffs developed at BASF.

1245. NIETZKI, R. Die Entwicklungsgeschichte der künstlichen organischen Farbstoffe. Stuttgart, 1902. (20 pp.)
A brief, undocumented but authoritative summary.

1246. PERKIN, W.H. "On colouring Matter Derived from Coal-tar," Quarterly Journal, Chemical Society (London), 14 (1861), 230-31.
An historical description of an industry that was five years old.

1247. PERKIN, W.H. "On the Analine or Coal Tar Colours," Journal, Society of Arts (London), 17 (1868-69), 99-105. (Reprinted in Gardner, no. 1234: 1-45)
A condensed history without citations. Illustrations of apparatus.

1248. PERKIN, W.H. "The History of Alizarine and Allied Colouring Matters, and Their Production from Coal Tar," Journal, Society of Arts (London), 27 (1878-79), 572-602.
A very detailed account of the manufacturing processes, with illustrations of apparatus. No citations.

1249. PERKIN, W.H. "The Origin of the Coal Tar Colour Industry and the Contribution of Hofmann and His Pupils," Journal, Society of Chemical Industry (London), 69 (1896), 596-637. (reprinted in Gardner, no 1234, 141-87)
Good circumstantial account with references in text.

1250. REISERT, Arnold. "Geschichte und Systematik der Indigosynthesis," ZAC (1898), 1026.
A brief report on an address and description of a book of the same title, which I have been unable to locate.

1251. ROBINSON, Robert, ed. Perkin Centenary, London: One Hundred Years of Synthetic Dyestuffs. London, 1958. (136 pp.)
Proceedings of a symposium. Aside from a life of Perkin, by John Read (1-36, with citations), the articles are of little historical value.

1252. ROSCOE, H.E. "The Artificial Production of Alizarine," in Gardner, no. 1234, 46-53.
Summary with a chronology. No citatons.

1253. ROWE, F.M. The Development of the Chemistry of Commercial Synthetic Dyes (1856-1938). London: Institute of Chemistry, 1938. (106 pp.)
An important technical account, historically organized, with telegraphic citations in the text.

1254. SCHUSTER, Curt, and Willibald ENDER. "Alizarine," Die BASF, 20 (1970), 195-208.
Summary with illustrations.

1255. SCHLENK, O. "Die Entdeckung des Anilins," ZAC, 39 (1926), 25 ff.
Not seen.

1256. SONNEMANN, Rolf. Zur Geschichte der Teerfarben-Industrie in Deutschland von ihren Anfängen bis zur Bildung der beiden Dreibünde (1905/07). Leuna-Merseburg, 1963. (80 pp.)
A dissertation for the Institut für Marxismus-Leninismus of the local Technische Hochschule. Interesting.

1257. VOEGLIN, W. et al. "Sir William Henry Perkin," Ciba Review, no. 115 (1956), 2-49.
A series of short pieces on the early dye industry, well-illustrated and in the attractive style characteristic of this publication.

1258. VOGEL, Max. Die Entwicklung der Anilin-Industrie. Leipzig, 1866. (160 pp.)
Comprehensive with extensive citations. The last third is concerned with practice and includes a short section on toxicology.

1259. WARD, E.R. "C.B. Mansfield vs. F. Grace Calvert: A Forgotten Controversey in the Coal Tar Industry," Chemistry and Industry, 1957, 159.
Mansfield patented in 1847-48 a "benzole-gas or air light apparatus." Calvert delivered a paper on the subject in 1854, leading to a priority dispute relating to the production methods of the raw materials of the coal tar dyes. Citations.

1260. WARD, E.R. "Politics and 19th Century Organic Chemical Technology," Journal, Society of Dyers and Colourists, 80 (1964), 252-55.

Figure 8. Electric battery, plating tank, and electrical connections of the galvanoplastic works at Auteuil, France. From Turgan (item 226)

An historian enters the debate among correspondents to the journal (including Cliffe, no. 1228) over the decline of the British dyestuffs industry.

1261. WELHAM, R.D. "The Early History of the Synthetic Dye Industry," Journal, Society of Dyers and Colourists, 79 (1963), 98-105, 146-52, 181-85, 229-37.
Part I details the chemistry and technology of the most important dyes to 1900. Part II, referring to Britain, analyzes the decline of the industry, which is attributed to German advantage resulting from the lack of a German patent law until 1876, the failure of the British industry to retain talented managers, and the tendency of British education to produce "artisans" rather than technologists with some scientific education. Citations.

1262. WHITE, Howard J., Jr., ed. Proceedings of the Perkin Centennial, 1856-1956. New York: American Assn. of Textile Chemists and Colourists, n.d. (1956) (467 pp.)
Includes the proceedings of a symposium, mostly not historical. A few important historical articles have been entered separately.

ELECTROCHEMICAL INDUSTRIES

HISTORIES

1263. ACHESON, E.G. "Graphite: Its Formation and Manufacture," Chemical News, 79 (1899), 290-93.
Preprint of a piece intended for JFI, describing his new graphite factory and giving a brief history of the material.

1264. CHRISTOFLE, Charles. Histoire de la dorure et de l'argenture électrochimiques, Paris, 1851. (446 pp.)
Not seen.

1265. DUBPERNELL, George, "The Development of Chromium Plating," Plating Magazine, 47 (1960), 35-53.
From the deposition of chromium from chromic acid in 1856 to the date of publication; a straightforward chronology with ample citation.

1266. HABER, Fritz. "On the Electrochemical Industry in the United States," Electrochemical Industry, 1 (1902-03), 349-51, 379-97, 471.
Reports of a sixteen-week visit in 1901, to which some comments by Americans are added.

1267. LE SUEUR, E.A. "Inception and Development of Electrolytic Alkali Manufacture," Transactions, Electrochemical Society, 63 (1933), 187-95.
Personal reminiscences of "the father of the electrolytic caustic industry in North America." No references.

1268. LUNGE, Georg. "Geschichte der Elektrolyse von Chloriden," ZAC, (1896), 517-20.
Comments on claims recently made for the priority of various persons (Hargreaves, Richardson, Castner) in recent publications. No citations.

1269. QETTEL, Felix. "Entwicklung der electrochemischen Industrie," SC, 1, Hft. 3 (1896), 89-120.
Straightforward narration from galvanoplastics in the 1830's to recent applications to organic materials. The principal obstacle is the cost of electricity. No citations.

1270. RICHARDS, Joseph W. "The Electrochemical Industries of Niagara Falls," Electrochemical Industry, 1 (1902-03), 11-23, 49-55.

1271. TRESCOTT, Martha M. The Rise of the American Electrochemicals Industry, 1880-1910. Studies in the American Technological Environment. Westport, Conn.: Greenwood, 1981. (391 pp.)
A hundred pages of straightforward history are followed by auxillary chapters, on Niagara Falls power, the unit operations concept, and economic, labor and educational aspects of the chemical industry. Much space is given to discussion of the views of other historians. Citations.

1272. VORCE, L.D. "Historic Development of Caustic-chlorine Cells in America," Transactions, Electrochemical Society, 86 (1944), 69-81.
A chronology with citations of sources.

EXPLOSIVES

HISTORIES

1273. AYERST, R.P., et al. "The Role of Chemical Engineering in Providing Propellants and Explosives for the U.K. Armed Forces," in Furter (1980, no. 116), 367-92.

1274. BUJARD, Alfons. Zündwaren. Leipzig, 1910. (130 pp. Sammlung Goschen)
Noteworthy principally for the citations included in its brief historical section (pp. 10-27).

1275. CRASS, M.F., Jr. "A History of the Match Industry," JCE, 18 (1941), 116-20, 277-82, 380-84, 428-31.
To about 1900, and mostly dealing with the U.S. Important for references to patents.

1276. DIXON, Wm. H. The Match Industry: Its Origins and Development. London, 1925.
Includes 70 pp. of history but semi-popular and without citations.

1277. DUTTON, Wm. S. One Thousand Years of Explosives. Philadelphia, 1960.
Trivial.

1278. FISCHER, Hugo. "Der Bickfordische Sicherheitszünder und die Errichtung der ersten Sicherheitzünderfabrik in Deutschland," Technikgeschichte, 6 (1914), 55-78.
Describes in detail the invention of "the miner's safety fuse" by Wm. Bickford of Cornwall in 1831, its manufacture in England and the first production of the device in Germany, near Meissen, in the early 1840's. Citations, especially to patents.

1279. GEORGE, Stephen L. "The Origins and Discovery of the First Nitrated Organic Explosives." Dissertation, Univ. of Wisconsin, 1978. (Univ. Microfilms 78-4417).
Discusses the work of T.J. Pelouze, C.F. Schönbein and Ascanio Sobrero, the introduction of mixed acid for nitration and the replacement of gunpowder by smokeless powder, which George dates from the 1890's. Numerous citations.

1280. HASSENSTEIN, W. "Der Uebergang vom Schwarzpulver zum Nitrozellulose-Blattschenpulver vor 50 Jahren," Zeitschrift für die Geschichte von Schiess- und Sprengstoffwesen, 36 (1941), 75-78, 100-103, 120-21. Not seen.

1281. HEAVISIDE, M. The True History of the Invention of the Lucifer Match, by John Walker, of Stockton-on-Tees, 1827. Stockton-on-Tees, 1900. (32 pp.)
Consists mainly of quotations from elderly residents of the town.

1282. HODGETTS, E.A.B. "The Rise and Progress of the British Explosives Industry," 7th International Congress of Applied Chemistry, London (1909), 3-39.
This brief sketch inspired the writing of Van Gelder's 1132 pp. history of the American industry (no. 1290)!

1283. MACDONALD, Georg W. Historical Papers on Modern Explosives. London, 1912. (192 pp.)
Twenty-seven short papers which had previously appeared in the periodical Guns and Explosives. Most are concerned with "gunscotton," from Schönbein (1846) to Nobel. Few citations.

1284. MARSHALL, Arthur. Explosives. London, 1915.
Contains a good summary history, pp. 1-43.

1285. ROMOCKI, J.J. von. Geschichte der Explosivstoffe. Berlin, 1895-96. (2 vols)
A standard work, now dated, but not replaced.

1286. SPETER, Max, "Die Detonierungsversuche Fourcroys-Vauquelise mit der Knallsalz," Zeitschrift für die Geschichte von Schiess- und Sprengstoffwesen, 27 (1932) 322 ff.
Not seen.

1287. TULLOCH, T.G., ed. The Rise and Progress of the British Explosives Industry. London: 7th International Congress of Applied Chemistry: Explosives Section, 1909. (417 pp.)
Historical chapters by various authorities with some citations and a substantial bibliography. Includes a chronology, 1242 to 1700, and a list of manufacturers to 1880.

1288. UPMANN, J. Schiesspulver, dessen Geschichte, Fabrikation, Eigenschaften und Proben. Braunschweig, 1874. (217 pp.)
Includes twelve pages of history without citations and largely outdated.

1289. URBANSKI, M.T. "Le centenaire de la nitrocellulose," Mémorial de l'Artillerie Française, 13 (1934), 825-41.
Quotations, from H. Braconnet's "De la transformation de plusiers substances végétales en principe nouveau" (AC, 52 [1833], 290 ff.) to sources on the development of stable powder in the 1860's. Numerous citations.

1290. VAN GELDER, A.P. and Hugo SCHLATTER. History of the Explosives Industry in America. New York, 1927. (1132 pp.)
Includes chronological histories of black powder, nitroglycerine and dynamite, fuses and blasting caps, interrupted by company histories, which account for the bulk of the volume. Citations.

1291. ZORN, Walther. Die deutsche Zündholzindustrie. Leipzig, 1913. (185 pp.)
A dissertation, devoting 100 pages to history, principally economic and social. Says that matches were brought before the public in 1833, from three sources, John Walker (Britain), Iriny (Hungary), and Johann F. Kammerer (Germany). Citations.

FUELS, EXOTIC

HISTORIES

1292. BIRKENFELD, Wolfgang. Der synthetische Triebstoff, 1933-45: ein Beitrag zur nationalsozialistischen Wirtschafts- und Rüstungspolitik. Göttingen, 1964. (280 pp.)
Primarily a study of war economics as illustrated by the German program for producing fuel oil by coal hydrogenation. Citations include many unpublished sources.

1293. CLARK, John D. Ignition! An Informal History of Liquid Rocket Propellants. New Brunswick, N.J.:

Rutgers Univ. Press, 1972. (214 pp.)
Informal to the degree that it lacks both citations and bibliography. But it gives a profuse account of the arcane research in rocket fuels, to the date of publication. The author was in charge of development at the U.S. Rocket Test Station, Dover, N.J.

1294. GIEBELHAUS, August W. "Farming for Fuel: The Alcohol Motor Fuel Movement of the 1930's," Agricultural History, 54 (1980), 173-84.
On the "power alcohol movement" in the U.S. in the 1930's.

1295. KRAMMER, Arnold. "Technology Transfer as War Booty: the United States Technical Oil Mission to Europe, 1945," T&C, 22 (1981), 68-103.
"Germany's substantial advances in numerous industries--especially rocketry, optics, plastics, industrial chemistry, pharmaceuticals and synthetic fuel technology--would be of substantial value to both American industry and to the continuing war effort against the Japanese." Here illustrated by synthetic petroleum. A thought-provoking account of victorious America's zeal for German technology and the consequences or, rather, the lack of consequences.

1296. NAGEL, Alfred von. Methanol. Triebstoff, Ludwigshafen, 1970. (79 pp., SB no. 5)
The history of high pressure synthesis and fuel hydrogenation at BASF.

1297. PIER, Mathias. "Einiges aus der Entwicklung der katalytischen Druckhydrierungen," Zeitschrift für Elektrochemie, 57 (1953), 456-60.
An address dealing mainly with the period 1913-53. Summary but important. The author was with BASF.

GAS, HEATING AND ILLUMINATING

EARLY WORKS

1298. ACCUM, Freidrich. Description of the Processes of Manufacturing Coal Gas, for the Lighting of Streets, Houses, and Public Buildings... London, 1815. (186 pp.)

Figure 9. Battery of retorts used in the production of gas for heating and illumination. From Turgan (item 226)

An omnium gatherum, ranging from meditations on the usefulness of the arts to instructions for workmen "attending the gas light apparatus." Includes some history, a price list of the apparatus, and colored prints, which have made it a "rare book."

HISTORIES

1299. AMERICAN GAS ASSOCIATION. Historical Statistics of the Gas Industry. New York, 1956.
Refers to the U.S. and goes back, in most cases, no more than about 25 years.

1300. BADER, Louis. "Gas Illumination of New York City, 1823-63." Diss., New York University, 1971 (Univ. Microfilm, 71-2261)
Not seen.

1301. BERTELSMANN, W. "Die Entwicklung der Leuchtgaserzeugung seit 1890," SC, 12, Hft. 7/8 (1907), 231-320.
A retrospective account of the state of the art. Few citations but numerous illustrations and tables.

1302. BLOCHMANN, Georg M.S. Beiträge zur Geschichte der Gasbeleuchtung. Dresden, 1871. (124 pp.)
The most complete work on this topic with profuse citation. Deals with England, France, Germany, and the United States, methods and apparatus, statistics, technology, and the influence of science.

1303. BROWNLIE, David. "The Early History of the Coal Gas Process," TNS, 3 (1922), 57-68.
Principally a summary description of the work of Wm. Murdoch, Philippe Lebon, and Wm. Bruton, from 1790 to 1820.

1304. CHANDLER, Dean. Outline of the History of Lighting by Gas. London: South Metropolitan Gas Co., 1936. (279 pp.)
Not seen.

1305. CLEGG, Samuel, Jr. A Practical Treatise on the Manufacture and Distribution of Coal Gas. London, 3rd ed., 1859. (394 pp.)

This edition begins with a 23-pp. "historical sketch," dating the subject from a note in the Philosophical Transactions for 1667 on "inflammable gas" discovered in a well. Stephen Hales referred to "elastic inflammable air" in his <u>Vegetable Statics</u> (1726) after which "coal gas" was a common "philosophical experiment." Says that Murdoch was lighting his house with gas by 1792, Lebon about the same time. Clegg, an important early figure in this technology, published the first edition in 1841 in 208 pp.

1306. CRACROFT, P.G. "The Early Days of Gas Lighting," <u>The Engineer</u>, 150 (1930), 616-17, 642-43.
From Paracelsus (!) to Murdoch but largely restricted to Britain. A summary account.

1307. ELTON, Arthur. "The Rise of the Gas Industry in England and France," <u>Acts, 6th International Congress of the History of Science</u> (Amsterdam), 1950. 2:492-504.
A "history of the application of theoretical principles to practice." Reviews the work of inventors, the ultimate demand--to enable textile factories to work at night--and gives the "palm" for success to Albert Winsor, founder of the Gas Light and Coke Co. (1812).

1308. HUNT, Charles. <u>A History of the Introduction of Gas Lighting</u>. London, 1907. (150 pp.)
Deals with Wm. Murdoch (1754-1839) and his predecessors. Citations. Illustrated.

1309. KOERTING, A. "Geschichte der Gastechnik," <u>Technikgeschichte</u>, 25 (1936), 84-110.
A history of the generation and application of gas for heating and lighting in Germany, relating this to its development in England. Bibliography but no citations.

1310. MACFARLAN, J. "George Dixon: Discoverer of Gas Light from Coal," <u>TNS</u>, 5 (1925), 53-55.
Modifies the claims made for Murdoch's work in 1810 by citing that of Dixon about 1759.

1311. MATTHEWS, Wm. <u>Historical Sketch of the Origin, Progress, and Present State of Gas Lighting</u>. London, 1827, (434 pp.)

Covers the period from Thos. Shipley's note in the
<u>Philosophical Transactions</u> for 1667 to Parliamentary
hearings on the Chartered Gas Works in the 1820's.
Numerous appendices including the personal accounts of
Murdoch and Clegg.

1312. PEEBLES, Malcolm W.H. <u>Evolution of the Gas Industry</u>.
New York: New York Univ. Press, 1980. (235 pp.)
The traditional "coal gas" industry is treated very
briefly as a prelude to a description of the natural
gas industry in the United Kingdom, United States,
Japan, the Netherlands and the USSR. Concludes with a
chapter on LNG (liquified natural gas). Brief bibliography, no citations.

1313. STORZ, Louis, and A. JAMISON. <u>History of the Gas
Industry</u>. New York, 1938. (542 pp.)
In the United States. Only the first 140 pp. are
historical and popular at that. No citations. Of
little value.

1314. SZEPCZYNSKI, S. de. "Zur Geschichte der Acetylenbeleuchtung," <u>Zeitschrift für Carbid und Acetylen</u>, 2
(1898), 242-44, 250-52, 258-59, 266-67.
Not seen.

1315. WILSON, Channing W. "Foundations and Development of
the Gas Industry in America," <u>JCE</u>, 18 (1941), 103-07.
Brief but circumstantial. No citations.

GASES, INDUSTRIAL

EARLY WORKS

1316. "Descriptions d'un nouvel appareil pour procurer à peu
de frais et avec abondance au blanchisseries le
chlor qui leur est nécessaire," <u>Annales de l'Industrie</u>, 8 (1822), 135-42.
Description of the "chemical cascade" of Berthollet,
Descroizilles and Clement for the production of
chlorine.

1317. BOUSSINGAULT, J.B.J.D. "Sur l'extraction du gaz oxygèn
de l'air atmospherique," <u>CR</u>, 32 (1851), 261-67, 811-13.

Based on the reversability of the oxygenation of barium monoxide to the dioxide. In an improved version this was the principal process for producing industrial oxygen as late as 1900.

HISTORIES

1318. "Wasser und Wasserstoffgas, und die Darstellung des Letztern zu technischen Zwecken," NN, 1 (1854), 48-58.
Review of recent literature on the commercial production of hydrogen.

1319. "Chlorine--a Key Material," CIBA Review, 12, no. 139 (1960), 1-35.
A popular but important summary of the history of the production and use of the material. Bibliography. Illustrated.

1320. BALDWIN, R.T. "History of the Chlorine Industry," JCE, 4 (1927), 313-19, 454-59.
Summary, the second part dealing with its uses, which were (to the producers) distressingly few. The author was Secretary of the "Chlorine Institute."

1321. BERNSTHEN, A. "Die synthetische Gewinnung des Ammoniaks," ZAC, 26 (1913), 10-16.
Historical. References.

1322. BRENEMANN, A.A. "The Fixation of Atmospheric Nitrogen," Journal, American Chemical Society, 11 (1889), 2-27, 30-48.
Historical orientation but deals almost exclusively with cyanogen. Citations.

1323. CLAUDE, Georges, "La recherche scientifique, ses applications à l'industrie et la synthèse industrielle de l'ammoniac," Revue Generale des Sciences, 32 (1921), 534-43, 570-81.
An address to the French Association for the Advancement of Science by the originator of the French process for synthetic ammonia. Rhetorical but informative.

1324. DOBBIN, Leonard, "The History of the Discovery of Phosgene," AS, 5 (1945), 270-87.
A detailed account. In that age of chemical innocence (c. 1810) no notice was taken of its possible use as a war gas.

1325. DUMONT, Georges, and E. HUBON. Historique, propriétes, fabrication, apparatus, de l'acetylene. Paris, 1896. (122 pp.)
Not seen.

1326. FRIES, A.A. "Use and importance of Chlorine in Chemical Warfare," Transactions, American Electrochemical Society, 49 (1926), 181-88.
After the first gas attack, April 22, 1915, "better" was gases were developed--but all contained chlorine.

1327. HARDIE, D.W.F. Acetylene, Manufacture and Uses. London, 1965.
It concludes (76-100) with an "Outline of historical development," and a chronology. No references but authoritative, the author being well known for his historical work.

1328. HENGLEIN, F.A. "150 Jahre Stickstoffdünger," Chemiker Zeitung, 75 (1951), 345-50, 389-92, 407-10.
Essentially a history of the Haber-Bosch ammonia synthesis, through biographies of Haber and Bosch. Some references.

1329. HOUSEMAN, C.R. "Industrial Oxygen Production, an Historical Survey," Proceedings, International Congress of Acetylene (London), 1936: 882-92.
Not seen.

1330. JEFFERY, T.P. et al., eds. Chlorine Bicentennial Symposium. Princeton: The Electrochemical Society, 1974.
A miscellany with some historical information.

1331. LANG, Arnold. Histoire du carbure de calcium et ses ententes internationales réglant la production et la vente de ce produit. Geneva, 1949. (329 pp.)
A useful description although largely devoted to business history and cartelization. A few citations.

1332. MITTASCH, Alwin. Geschichte der Ammoniaksynthese.
Weinheim: Verlag Chemie, 1951. (196 pp.)
An account, both scholarly and personal, of perhaps
the most momentous event--prior to 1945--in the history of industrial chemistry. The author was an
important participant (I have seen his letter to an
American friend, written in the late 1940's, regretting that the project had been successful.)

1333. MOND, Ludwig, "History of the Manufacture of Chlorine,"
Journal, Society of Chemical Industry, 15 (1896),
713-19.
Presidential address to the BAAS. Summary but with
considerable detail. No citations.

1334. MOORE, R.B. "Helium: Its History, Properties, and
Commercial Development," JFI, 191 (1921), 145-97.
A full treatment with much historical information.
Citations.

1335. MURRAY, R.L. "The Chlor-alkali Industry in the United
States," I&EC, 41 (1949), 2155-64.
There was a large jump in chlorine production about
1917, a smaller one in 1940. These are mainly economic ruminations about a material for which demand
still lagged behind supply.

1336. NAGEL, Alfred von. Stickstoff, Ludwigshafen, 1970.
(SB no. 3)
A history of nitrogen fixation by high pressure synthesis at BASF.

1337. PICTET, R. L'acetylene, son passé, son présent, son
avenir. Lyon, 1896.
Not seen.

1338. PRITCHARD, D.A. "Economics of Chlorine," Transactions,
American Electrochemical Society, 49 (1926), 29-41.
Mainly a history of methods of manufacture and discussion of the problems of what to do with excess production.

1339. SIMPSON, Charles H. Chemicals from the Atmosphere.
New York, 1969. (181 pp.)
A well-written popular account. No citations.

1340. TIMM, Bernhard. "50 Jahre Ammoniak-Synthese," <u>Chemie-Ingenieur Technik</u>, 35 (1963), 817-23.
A brief but reliable summary of the technical consequences of the establishment of a synthetic ammonia works by BASF on Sept. 9, 1913.

1341. VOGEL, J.H. <u>Handbuch für Acetylen</u>. Braunschweig, 1904. (880 pp.)
History of calcium carbide, 3-13; history of acetylene, 127-36. Annotated chronologies with citations. Credits Thos. L. Willson, at Spray, North Carolina, with the first commercial exploitation of carbide, in 1895.

1342. WAESER, Bruno. <u>The Atmospheric Nitrogen Industry</u>. London, 1926. (746 pp. trans. from the German ed. of 1922)
A detailed description of the development of nitrogen fixation by all processes and in all countries. Written at a time when the industry was little more than a decade old.

1343. WHITTEMORE, Gilbert, F., Jr. "World War I, poison gas research, and the ideals of American chemists," <u>Social Studies of Science</u>, 5 (1975), 135-63.
Not seen.

INORGANIC SYNTHESIS

HISTORIES

1344. "Das Wasserglas. Erfindung, Wesen, und Nutzen derselben," <u>NN</u>, 7 (1857), 10-25.
Attributes the discovery of waterglass (soluble sodium silicate) to Oberbergrath Fuch, of Munich. This article is followed by another, "Das Wasserglas in der Industrie," (Ibid., 8 [1858] 1-43).

1345. BULLIER, L.M. "Geschichte des Calcium-carbids," (<u>Zeitschrift für) Carbid und Acetylene</u>, 1 (1897), 1-2.
Not seen.

1346. LUDWIG, A. "Der Entdecker des Calciumcarbids," (<u>Zeitschrift für) Carbid und Acetylene</u>, 2 (1898), 49-50.

Not seen. Reportedly attributes the discovery to Robert Hare.

1347. LUDWIG, A. Führer durch die gesammte Calciumcarbid literatur. Berlin, 1899.
Lists (p. 16) the earliest publications on carbide.

1348. MANTIGNON, Camille. "Le centenaire de l'iode: les conséquences de sa découverte," Revue Générale de Chimie, 16 (1913), 391-99.
Industrial consequences. References. Important.

1349. MANTIGNON, Camille. "L'Industrie de l'iode, son histoire..." Revue Générale des Sciences, 25 (1914), 511-16.
Dates the industry from the 1830's. Citations and production data.

1350. TAUSSIG, Rudolf. Die Industrie des Kalziumkarbides. Halle, 1930. (518 pp.)
History, 3-23, emphasizing the period after 1895 (Vogel, no. 1341, emphasizes the earlier period). Details the work of Henri Moisson and Thos. L. Willson. Illustrations and references, largely to patents.

NITRATES

HISTORIES

1351. BERNTHSEN, A. "Ueber Luftsalpetersäure," ZAC, 22 (1909), 1167-78.
Historical, focusing on the works at Kristianssand, Norway, with important illustrations of apparatus. Citations.

1352. CHILTON, T.H. Strong Water: Nitric Acid, Its Sources, Methods of Manufacture, and Uses. Cambridge, Mass., 1968. (170 pp.)
History, "from the alchemists through World War I," is covered in the first 24 pages but the remainder is written with an awareness of history unusual in the profession today (the author was a retired Du Pont "engineering department scientist"). The book appears to be an attempt to give students of chemistry or en-

gineering some feeling for the context of their specialties, and is to be commended for that. Few notes and no bibliography.

1353. CLARKE, M.J. "The Federal Government and the Fixed Nitrogen Industry, 1915-1926." Dissertation, Oregon State Univ., 1977.
Not seen.

1354. GROSSMANN, H. "Stickstoffindustrie und Weltwirtschaft. SC, 28, Hft. 8/9 (1924-26), 255-346.
Historical orientation.

1355. GROSSMANN H., ed. Der Kalkstickstoff in Wissenschaft, Technik, und Wirtschaft. SC, n.s. 6 (1931). (213 pp.)
Includes F. Janke, "Geschichte, Technik und Wirtschaft des Kalkstickstoffs," Heinrich Franck, "Chemie des Kalkstickstoffs," and Willy Makkus, "Das Düngmittel Kalkstickstoff." Each begins with a significant historical introduction. References.

1356. HOECHST (firm). Wilhelm Ostwald und die Stickstoffgewinnung aus der Luft. Frankfurt/M, 1964.
(DHA no. 5)
Not seen.

1357. MITTASCH, Alwin. Saltpetersäure aus Ammoniak. Weinheim, 1953. (136 pp.)
Scientific and technical history to 1920. The author was an important participant in high-pressure catalytic research at BASF. Citations. Important.

1358. PERLICK, Anton F.W. Der derzeitige Stand der Luftstickstoff-Industrie. Heidelberg, 1913. (26 pp.)
A dissertation describing the development and consequences of the Birkland-Eyde process for nitrogen fixation. No references.

1359. SCHOENHERR, O. "Ueber die Fabrikation der Luftsalpeters nach dem Verfahren der Badischen Anilin- & Soda-Fabrik," Elektro-Technische Zeitschrift, 30 (1909), 397-402.
An account of laboratory experimentation at BASF, Ludwigshafen, and development work at Kristiansand, Norway.

1360. WAR, E.R. "Industrial Mixed Acid Nitration," Ambix, 23 (1976), 199-200.
The use of sulphuric acid in nitration was first patented by C.B. Mansfield, in England, 1847.

ORGANIC SYNTHESIS

HISTORIES

1361. AMERICAN CHEMICAL SOCIETY, DIVISION OF PETROLEUM. Symposium on Halogenation of Hydrocarbons and Properties of Products. St. Louis, 1948. (86 pp.)
Nine articles on various aspects of the topic, which was so new that 95% of the references in the introduction (Earl T. McBee, "Progress in chlorination") are dated 1940 or later. Under the circumstances, history is unavoidable. Although marked "not for publication," the book has found its way onto the shelves of the Library of Congress. Citations.

1362. HEUCK, Klaus. "Ein Beytrag zur Geschichte der Kautschuk-Synthese." Chemiker Zeitung, 94 (1970), 147-57.
Surveys the manufacture of methyl rubber, introduced 1906-18, and the history of BUNA rubber, 1926-45. Emphasizes the cooperation of individual chemists and firms, especially IG Farbenindustrie. Based largely on patents.

1363. ICI. Chapters in the Development of Industrial Organic Chemistry. London: ICI, n.d. (1937 ?) (112 pp.)
The "chapters" are apparently exhibits, here catalogued, celebrating the opening of a new research laboratory. Of no historical significance.

1364. KONRAD, E. "Zur Entwicklung des synthetischen Kautschuke in Deutschland," ZAC, 49 (1936), 799-801.
On the work of Fritz Hofmann, whose basic patent on the polymerization of isoprene initiated work at the firm of Bayer in 1907.

1365. NAGEL, Alfred von. Aethylen. Acetylen. Ludwigshafen, 1971. (74 pp., SB no. 7)
On the history of the exploitation of these raw materials, especially at BASF.

1366. PUMMERER, Rudolf. "Geschicte und Bedeutung des künstlichen Kautschuks," Technikgeschichte, 27 (1938), 134-51.
A brief but authoritative account, from the discovery of isoprene. Discusses diene, butadiene, buna and neoprene rubbers.

1367. SCONCE, J.S., ed. Chlorine. New York, 1962.
Not an historical work, but it inevitably deals with the history of the uses of chlorine in organic synthesis, for most of it was little over two decades old at the time of publication.

1368. SPETER, Max. "Berzelius' Katalysepublikationen," Chemiker Zeitung, 56 (1932), 561.
Lists relevant references from the celebrated chemist, who not only had an eye for application but had for a time his own chemical factory.

1369. WENDLAND, Ray T. Petrochemicals: The New World of Synthetics. New York, 1969. (299 pp.)
A popular introduction to the subject with historical asides, mostly without indication of where they came from.

PETROLEUM

HISTORIES

1370. CLARK, J.A. The Chronological History of the Petroleum and Natural Gas Industries. Houston, n.d. (1963 ?) (317 pp.)
From 6000 B.C. (when Noah smeared bitumen on the ark) to 1962. An idiosyncratic but honest book, which gives its sources--most of them secondary when not tertiary. Almost exclusively concerned (Noah notwithstanding) with the United States.

1371. ENOS, John L. Petroleum Progress and Profits: A History of Process Innovation. Cambridge, Mass., 1962
Deals with the several processes for "cracking" petroleum, from 1913 to the late 1930's. A profound technical and economic study. Citations.

1372. GIDDINS, P.H. The Birth of the Petroleum Industry. New York, 1938. (206 pp.)
A superior history, with ample citation, but almost nothing on petro-chemistry.

1373. FORBES, R.J. Studies in Early Petroleum History. Leyden, 1958. (199 pp.)
Thirteen articles on various aspects of the subject. Profuse citation. It was followed by More Studies... 1860-1880, Leyden, 1959 (199 pp.).

1374. HAYES, S. Dana. "On the History and Manufacture of Petroleum Products," American Chemist, 2 (1872), 401-05.
A brief sketch, ending with a description of the state of the art. Considers Joshua Merrill of the Downer Co., Boston, the most important figure. The author was state assayer of Massachusetts.

1375. JOHNS, Carlo. "The History and Status of Chemistry in Petroleum Research," I&EC, 15 (1923), 446-49.
Summary. No citations.

1376. LAWRENCE, Albert A. Petroleum Becomes of Age. Tulsa, 1938. (227 pp.)
A narrative history, restricted to the United States. Notable for its citations, especially to contemporary newspapers.

1377. MOSELY, Charles G. "The Capitalist, the Chemist, and Lima Sour Crude Oil," JCE, 56 (1979), 657-58.
The capitalist was J.P. Rockefeller, the chemist Hermann Frasch, and the oil a sulphur-containing crude oil from an Ohio well.

1378. RABKIN, Y.M. "La chimie et le pétrol: les débuts d'une liaison," Revue d'Histoire des Sciences, 30 (1977), 303-36.
Deals with the period 1850-80 and the limitations which inhibited the application of science to the petroleum industry in the great producing nations, the United States and Russia. Concludes that the study of the chemistry of petroleum was largely a product of the curiosity of emminent chemists of the period, especially in France; and that it had little connection with the industry. Citations.

1379. RABKIN, Y.M. "Chemicalization of petroleum refining in the United States: the role of cooperative research, 1920-1950," Social Science Information, 19, 4/5 (1980), 833-50.
Cooperative research--involving several institutions, industries, universities, or government--began in the 1920's, being promoted by the American Petroleum Institute. Citations.

1380. SCHMITZ, Pierre M.E. L'Epopée du petrol. Paris, 1947. (226 pp.)
Popular, but on a high level, with a full treatment of chemical aspects. Essentially ends at 1918. No citations.

1381. SEDILLOT, René. Histoire du petrol. Paris: Fayard, 1974. (362 pp.)
Up to date but considers petroleum only as fuel, virtually ignoring petrochemicals. No citations.

1382. SUHLING, Lothar. Erdöl und Erdölprodukte in der Geschichte. Munich: Deutsches Museum, 1975. (112 pp. Abhandlung und Berichte der Deutsches Museum, vol. 43, nos. 2 & 3)
From antiquity until "the beginning of large scale industrial production" (end of the nineteenth century). Narrative, with numerous citations. Illustrated.

1383. WILLIAMSON, Harold F. The American Petroleum Industry, 1859-1959. Evanston, Ill., 1959-63. (2 vols)
Although primarily an economic history, this well-written and documented book deals well, if succinctly, with technology. Petro-chemicals are treated very briefly.

PHARMACEUTICALS

HISTORIES

1384. BALDRY, P.E. The Battle against Bacteria: A History of Antibacterial Drugs. Cambridge, 1965. (102 pp.)
Popular summary without notes or bibliography.

1385. BELL, Jacob, and Theophilus REDWOOD. <u>Historical Sketch of the Progress of Pharmacy in Great Britain.</u> London, 1880. (415 pp.)
Pt. 1, by Bell and written in 1842, is on "early but unsuccessful attempts to separate pharmacy from medical practice" in Britain. Pt. 2 deals with the subsequent attempts leading to the "Pharmacy Act" of 1868, professionalizing pharmacy through examinations. Few references but much quotation of documents.

1386. COMROE, J.H. "Pay Dirt: The Story of Streptomycin," <u>American Review of Respiratory Diseases</u>, 117 (1978), 773-81, 957-68.
On the work of Selmin Waksman, 1939-44, and the application of streptomycin to tuberculosis by W.H. Feldman and H.C. Hinshaw. Citations.

1387. DE HAEN, P. "Golden Years of Drug Introduction, 1941 to 1970," <u>New York State Journal of Medicine</u>, 72 (1972), 253-58.
New "single drugs" marketed in the United States numbered 917--"there will never be a time like this again." Discusses them in terms of application. No citations.

1388. EICHGRUEN, A. "25 Jahre Arzneimittel Synthese," <u>ZAC</u>, 26 (1913), 49-54.
An address on the 25th anniversary, both of the Verein Deutscher Chemiker and of the deliberate (planned) synthesis of drugs by Carl Duisberg and Otto Hinsberg at the Elberfeld Dyeworks. Circumstantial without references.

1389. ELDER, Albert L., ed. <u>The History of Penicillan Production</u>. New York: American Institute of Chemical Engineers, 1970. (100 pp.)
Technical history, 12 chapters by 16 authors, many of whom were participants. Citations. This is an unusually successful product of a "one day history session" at a professional meeting.

1390. ERNST, Elmar. <u>Das "Industrielle" Geheimmittel und seine Werbung</u>. Würzberg: JAL Verlag, 1975. (373 pp., Quelle und Studien zur Geschichte der Pharmazie, Bd. 12)

On large-scale production of proprietary medicines in Germany in the second half of the nineteenth century. Citations.

1391. FELDMAN, Harry A. "The Beginning of Antimicrobial Therapy: Introduction of the Sulfonamides and Pencillans," Journal of Infectious Diseases, 125 (1972), Supplement, 22-25.
Random but interesting information. Sulphonamide was brought forward as a dye in 1908, its bacteriacidal properties were noted in 1913. Gerhard Domagk patented it in 1932 for its anti-strepticocal action (for which he won the Nobel Prize for 1939).

1392. GARROD, L.P., et al. "Evolution of the Anti-microbic Drugs," Antibiotics and Chemotherapy (Edinburgh, 4th ed., 1973) 1-11.
Summary.

1393. GUERRA, F. "Drugs from the Indies and the Political Economy of the 16th Century," Analecta medico-historica, 1 (1966), 29-54.
Miscellaneous information with numerous useful references. Notes that the terms drugs and spices were virtually interchangeable (to which I would add perfumes and dyes).

1394. HANGARTER, W. "Die Evolution eines Jahrhundret-Pharmakons-Acetylsalicylsäure--für die Medizin," Mediinische Welt, 25 (1974), 1968-78.
Some early information but chiefly concerned with developments after the introduction of "asperin" in 1899 by Farbfabrik vonm. Friedrich Beyer & Co. Citations.

1395. HARE, Ronald. The Birth of Penicillan and the Disarming of Microbes. London, 1970. (236 pp.)
Not seen. The author is particularly known for his work in the development of sulfa drugs.

1396. HOECHST, firm. Dr. Friedrich Stolz, der Erfinder des Pyramidons. Frankfurt/M, 1965. (64 pp., DHA no. 12)
Prints letters, patents, etc., from 1886, relating to the development of the first synthetic hormone.

1397. HOECHST, firm. Vorarbeiten zum Salvarsan. Frankfurt/M, 1966. (100 pp. DHA no. 14)
Letters, chiefly 1909-11, mainly from Paul Ehrlich. Continued in DHA no. 19 (Um die Zubereitung des Salvarsan. 1966. 87 pp.).

1398. HOECHST, firm. Die Zusammenarbeit Behring - Hoechst, 1892-1904. Frankfurt/M, 1968 (DHA no. 37)
Documents relative to the work of Emil Behring on serum therapy.

1399. HUBBARD, W.N. "The Origins of Medicinals," in John Z. Bowers and E.F. Purcell, eds. Advances in American Medicine, 2 (1978), 685-721.
From "the early colonial period" to the present, a useful summary with references.

1400. ISSEKULZ, Bela. Die Geschichte der Arzneimittel Forschung. Budapest: Akadémiai Kiado, 1971. (651 pp.)
A comprehensive history of pharmaceutical substances with a chronology up to 1964, listing them with citations to early references. Very valuable.

1401. KAUFMANN, G.B, and P.M. PRIEBE, "The discovery of saccharine," Ambix, 25 (1978) 191-207.
The discovery was "an integral part of (Ira) Remsen's research programme;" but it "might not have ever been exploited commercially" without "the expert knowledge of sugar chemistry" of (Constantin) Fahlberg. Thus they may be regarded as co-discoverers. Fahlberg patented saccharin and began its manufacture in New York in 1884 and in Leipzig in 1886. Citations.

1402. KENDALL, E.C. Cortisone. New York: Scribner, 1971. (175 pp.)
A charming and illuminating autobiography by one who shared the Nobel Prize in 1950, for the development of cortisone. No references.

1403. KREMERS, Edward, and George URDANG. History of pharmacy. A guide and a survey. Philadelphia, 1941. (475 pp.)
The standard history in English, although very much a summary. The 4th ed., by Glenn Sonnedecker (Phila.: Lippincott, 1976. 571 pp.) devotes nearly three-quarters of its space to the United States. References.

1404. LIEBENAU, J.M. Medical science and medical industry, 1890-1929: a study of pharmaceutical manufacturing in Philadelphia. Diss., Univ. of Pennsylvania, 1981. (489 pp.)
Claims, and largely documents, that "medical manufacturing" assumed its modern form, 1890-1930. Based primarily on the study of firms in Philadelphia, "the major center of the American pharmaceutical industry." Citations.

1405. MACLEOD, J.J.R. "History of the Research Leading to the Discovery of Insulin," Bulletin of the History of Medicine, 52 (1978), 295-312.
An account written 1922-23, immediately after the discovery and first attempts at commercial production. The author shared with Frederick Banting the Nobel Prize (1923) for this work. Documents. No citations.

1406. MATHEWS, L.G. History of Pharmacy in Britain. London, 1962.
A standard history, topically organized, with references. Comes up to date although limited by its concern with a single nation.

1407. MIALL, L.M. "Historical Development of the Fungal Pharmaceutical Industry," in J.E. Smith and D.R. Berry, eds. The Filamentous Fungi, Vol. 1, New York: Wiley, 1975: 104-21.
An authoritative summary, with references.

1408. PATZER, H., et al. Johann Bartholomäus Trommsdorff (1770-1837) und die Begrundung der modernen Pharmazie. Leipzig, 1972. (295 pp.)
Not seen.

1409. PELT, Jean M. "L'étrange histoire des médicaments," in her Les médicaments, Paris: du Seuil, 1969, pp. 15-44.
An interesting summary from primitive medicines to the end of the eighteenth century. No references.

1410. RATH, U. Zur Geschichte der pharmazeutischen Mineralogie. Braunschweig, 1971. (273 pp.)
Not seen.

1411. REUTTER de ROSEMONT, L. Histoire de la pharmacie a travers les ages. Paris, 1931. (2 vols)
Contains brief but not altogether negligible references to manufacturing pharmacy. Few citations.

1412. RICHARDS, A.N. "Production of Penicillan in the U.S., 1941-46," Nature, 201 (1964), 441-45.
A narrative chronology. References.

1413. RODNAN, G.P., and T.G. BENEDIK, "The Early History of Antirheumatic Drugs," Arthritis and Rheumatism, 13 (1970), 145-65.
Deals with "venerable remedies," colchseine (from colchicum autumnale, autumn crocus) which was known from the sixth century, salicylates (from willow bark), known to Dioscorides and synthesized and commercialized by H. Kolbe and E. Lautemann in 1874, quinoline (1879), antipyrim (1884), acctznilid (1886), aminopyrene (1896), all of them synthetic drugs. Also discusses gold salts and X-ray therapy.

1414. SCHADEWALDT, H. "Vom Apothekenlaboratorium zur pharmazeutischen Grossindustrie," Forschungen. Praxis. Forbildung, 17 (1964), 512-19.
A capsule history of commercial drugs, a splendid example of what can be accomplished in a short article. Citations.

1415. SCHADEWALDT, H. "Die Entdeckung der Sulfonamid," Deutsche Medizinische Wochenscrift, 100 (1975), 2617-21.
Not seen.

1416. SCHELENZ, Herman. Geschichte der Pharmazie. Berlin, 1904. (934 pp.)
The first two-thirds is devoted to annotated descriptions, in rigidly chronological order, of events relating to pharmacy from 2000 B.C. to the end of the eighteenth century. The last third, "independent pharmacy," is a potpouri of technological, educational, and cultural matters. A valuable book.

1417. SCHNEIDER, Wolfgang. "Geschichte der deutschen chemisch-pharmazeutischen Industrie. Betrachtung zu die Werk von Wilhelm Vershofen," Die pharmazeutisch

Industrie, 20 (1958), 229-34.
Praise of Vershofen's book (no. 146) with a summary of the contents.

1418. SCHNEIDER, Wolfgang. "Der Weg von der Entdeckung bis zur Synthese des Morphin," Mitteilungen Deutscher Pharmaciegeschichte, 35 (1965), 85-90.
A summary, whose high points are F.W. Sertürner's separation of morphine from opium in 1806, the quantitative analysis of P.J. Pelletier and J.B. Dumas in 1825, evolution of a structural formula, 1880, and approaches to a synthesis, 1939-52. Bibliography.

1419. SCHNEIDER, Wolfgang. Geschichte der pharmazeutischen Chemie. Weinheim, Verlag Chemie, 1972. (376 pp.)
An outstanding history with citations representing something of a summary of the numerous publications of this prolific author. Especially noteworthy for his innovations in subdividing the topic into periods characterized by drug types.

1420. SCHROEDER, Winfried. Die pharmazeutisch-chemischen Produkte deutscher Apotheken zu Beginn des naturwissenschaftlich-industriellen Zeitalters. Braunschweig, 1960. (238 pp.)
A thorough and documented analysis of drugs common in Germany, 1800-70.

1421. SNOW, John. Chloroform and Other Anesthetics. London, 1858. (443 pp.)
Good historical introduction, pp. 1-24.

1422. SOCIETY OF APOTHECARIES (London). The Origin, Progress, and Present State of Various Establishments for Conducting Chemical Processes and Other Medicinal Preparations at Apothecaries Hall. London, 1823. (21 pp.)
The Society was authorized to regulate drug production in 1617 and opened a laboratory in its Hall in 1671. A plate by W.T. Brand shows its appearance at date of publication.

1423. SONNEDECKER, Glenn. "The Rise of Drug Manufacture in America," Emory University Quarterly, 21 (1965), 73-87.
Summary. A few references.

1424. STANDER, S. "Transatlantic Trade in Pharmaceuticals during the Industrial Revolution," Bulletin of the History of Medicine, 43 (1969), 326-43.
Gives the contents of shipments, mostly to the West Indies, by a single English firm, during the years before and after 1776.

1425. STIEB, E.W. Drug Adulteration: Detection and Control in 19th Century Britain. Madison, Wisc., 1966. (335 pp.)
Excellent description but with little analysis. Citations.

1426. WAKSMAN, S.A. The Antibiotic Era: A History of the Antibiotics. Tokyo: Waksman Foundation of Japan, 1975. (224 pp.)
A straightforward narrative by an author who won the Nobel Prize for related work in 1952. He dates the "beginnings of the antibiotic era" 1940-45.

1427. WIETSCHORECK, Herbert. Die Pharmazeutisch-Chemischen Produkte deutscher Apotheken im Zeitalter der Nachchemiatrie. Braunschweig, 1962. (367 pp., a publication of the Pharmaziegeschichtlichen Seminar, Technische Hochschule, Braunsweig)
A systematic inquiry into early nineteenth century drugs, with a view to their classification as traditional (iatrochemical) or new (scientific). Citations.

1428. WILSON, David. In Search of Penicillan. New York: Knopf, 1976 (298 pp.)
A detailed history with references.

1429. WUEST, H.M. "A Hundred Years of Alkaloid Industry," Chemistry and Industry, 56 (1937), 1084-92.
An illustrated summary, without references.

PHOTOGRAPHY

EARLY WORKS

1430. DAGUERRE, L.J.M. Historique et déscription des procédés du daguerréotype et du diorama. Paris, 1839. (79 pp.)

Mostly documents relating to the discovery. On the history of this publication, see P.G. Harmant, "Daguerre's manual: a bibliographic eniga," <u>History of Photography</u>, 1 (1977), 79-83.

HISTORIES

1431. BAIER, Wolfgang. <u>A Source Book of Photographic History / Quellen & Darstellungen zur Geschiche der Fotographie.</u> Halle, n.d. (1963 ?) (703 pp., 313 illus.)
Quotations are connected by comment, giving a continuous narrative, ending with "the beginning of a new era, c. 1925." Only the title and the table of contents are in English, all quotations being in or translated into German. Emphasizes German work. Citations.

1432. COE, Brian. <u>Color Photography: The First Hundred Years, 1840-1940.</u> London: Ash & Grant, 1978. (144 pp.)
A popular, illustrated history without citations, but authoritative. The author is curator of the British Kodak Museum.

1433. CORNWALL-CLYNE, Adrian. <u>Color Cimematography</u>, London, 1951. (780 pp.)
Begins with an "historical summary" (3-38), otherwise a chronology with citations. First published in 1936.

1434. EDER, Josef M. <u>History of Photography.</u> New York, 1945. (860 pp. Reprinted, New York: Dover, 1978)
Eder (1855-1944), a chemist, published the first part of this in 1881 and kept enlarging it. Its first title, <u>Geschichte der Photochemie</u>, indicates its orientation, which is still evident in the last edition. He begins with Aristotle, and goes on to produce a veritable goldmine of fact, opinion and citation. The indispensible book.

1435. FRIEDMAN, Joseph S. <u>History of Color Photography.</u> Boston, 1944. (514 pp.)

He notes the radical advances since Wall's book of 1925 (no. 1452), but in spite of the title (and in contrast to Wall's book) this reads like a scientific-technical manual with historical asides and references buried in the text. A new edition (1968) has 565 pp. and a bibliography.

1436. GERNSHEIM, H., and A. GERNSHEIM. The History of Photography. London, 1955. (395 pp. and 359 illus.)
To the late nineteenth century. References. This is an authoritative work, which appeared in a second ed. in 1969.

1437. HARMANT, P.G. "Paleophotographic Studies: Was Photography Born in the 18th Century?" History of Photography, 4 (1980), 89-95.
Concerning correspondence of Nicéphore Niépce, which seems to indicate photographic work at Cagliari, Sardinia, where he served in Napoleonic army in 1793.

1438. HOECHST, firm. Farbstoffe für die Photographie. Frankfurt/M, 1965. (77 pp., DHA no. 11)
Documents, 1903-36, on the development of color photography at Hoechst.

1439. HUNT, Robert. Researches on Light. London, 1844. (301 pp.)
Ch. 1 (1-36) gives a detailed account, chronologically organized, of relevant research up to Daguerre and Fox Talbot. References in text.

1440. JENKINS, Reese. Images & Enterprise: Technology and the American Photographic Industry. Baltimore: Johns Hopkins Press, 1975. (371 pp.)
A comprehensive technological and business history with extensive citation. Covers the period 1839-1925.

1441. MEES, C.E.K. From Dry Plates to Ektachrome Film. New York, 1961. (302 pp.)
"A story of photographic research," primarily at Eastman Kodak. The author was the first (1912) and long-time head of the Kodak research laboratory. No references.

1442. OSTROFF, Eugene, "Talbot's Earliest Extant Print, June 20, 1835, Rediscovered," Photographic Science and Engineering, 10 (1966), 350-54.
Describes a dated image received by the Smithsonian Institution from the heirs of W.H. Fox Talbot and the methods used in its identification. The author had discussed restoration of other Fox Talbot photographs in the Smithsonian in an earlier article, "Early Fox Talbot Photographs and Restoration by Neutron Irradiation," Journal of Photographic Science, 13 (1965), 213-27.

1443. OSTROFF, Eugene. "History of Photomechanical Reproduction," Journal of Photographic Science, 17 (1969), 65-80 ("Etching, engraving and photography") 101-15 ("Photography and photogravure").
The first article begins with Nicephore Niépce's introduction of heliogravure in 1826, the second with W.H. Fox Talbot's invention of contact-screen photogravure in 1852. Citations.

1444. OSTROFF, Eugene, and T.H. JAMES. "Gelatin Silver Halide Emulsion: A History," Journal of Photographic Science, 20 (1972), 146-48.
Argues that this photographic emulsion was described nearly two decades earlier than has been supposed by Robert Bingham in 1850. Reports tests verifying Bingham's description.

1445. OSTROFF, Eugene. "Herschel and Talbot: Photographic Research," Journal of Photographic Science, 27 (1979), 73-80.
Describes Fox Talbot's correspondence with John Herschel in 1839, stemming from the former's concern over the recently published Daguerre process. Citations.

1446. POTONNIEE, Georges. Histoire de la découverte de la photographie. Paris, 1925. (319 pp.)
To the end of the "daguerrean period," ca. 1851. Citations. There is an English translation (New York, 1936), which omitted the photographs, for technical reasons. They are indeed not very clear in the French edition but valuable, considering the subject.

1447. SCHAAF, Larry, "Herschel, Talbot and Photography: Spring, 1831 and Spring, 1834," History of Photography, 4 (1980), 181-204.
Another account of the Talbot-Herschel "flirtation" with photography, chiefly from letters in the Royal Society of London. Citations.

1448. SIPLEY, L.W. A Half-Century of Color. New York, 1951. (216 pp.)
A popular "compilation" without references (or index), based on correspondence, most of it with firms. The author was director of the American Museum of Photography (Philadelphia).

1449. SPETER, Max. "Daguerres verschollenen Verfahren von 1826 zur Erzielung von lichtempfindlichen Chlorsilber-Papier aus Chloräthyl und Silbernitrat," ZAC, 49 (1936), 238-39.
Concludes that Daguerre was "an original investigator and discoverer."

1450. STENGER, Erich. The History of Photography. Easton, Penna., 1939. (204 pp.)
Begins in 1727 with Johann H. Schulze, "who had no conception of the idea of photography." Notable for sections dealing with such topics as the origin of technical terms, the earliest supply houses, societies, periodicals, etc. Few references, compared to other histories of photography. This appears to be an expanded version of a book published in Germany about 1929, which had only 44 pp.

1451. THOMAS, David B. The First Negatives. An Account of the Discovery and Early Use of the Negative-Positive Photographic Process. London, 1964. (36 pp.)
An account and discovery and early use of the negative-positive photographic process.

1452. WALL, E.J. The History of Three-Color Photography. Boston, 1925. (747 pp.)
Downplaying priority claims, the author of this very thorough work makes it appear that color photography emerged in the last half of the nineteenth century from a largely scientific matrix. So far as its commercial application was concerned, it was still emerging in 1925. Citations.

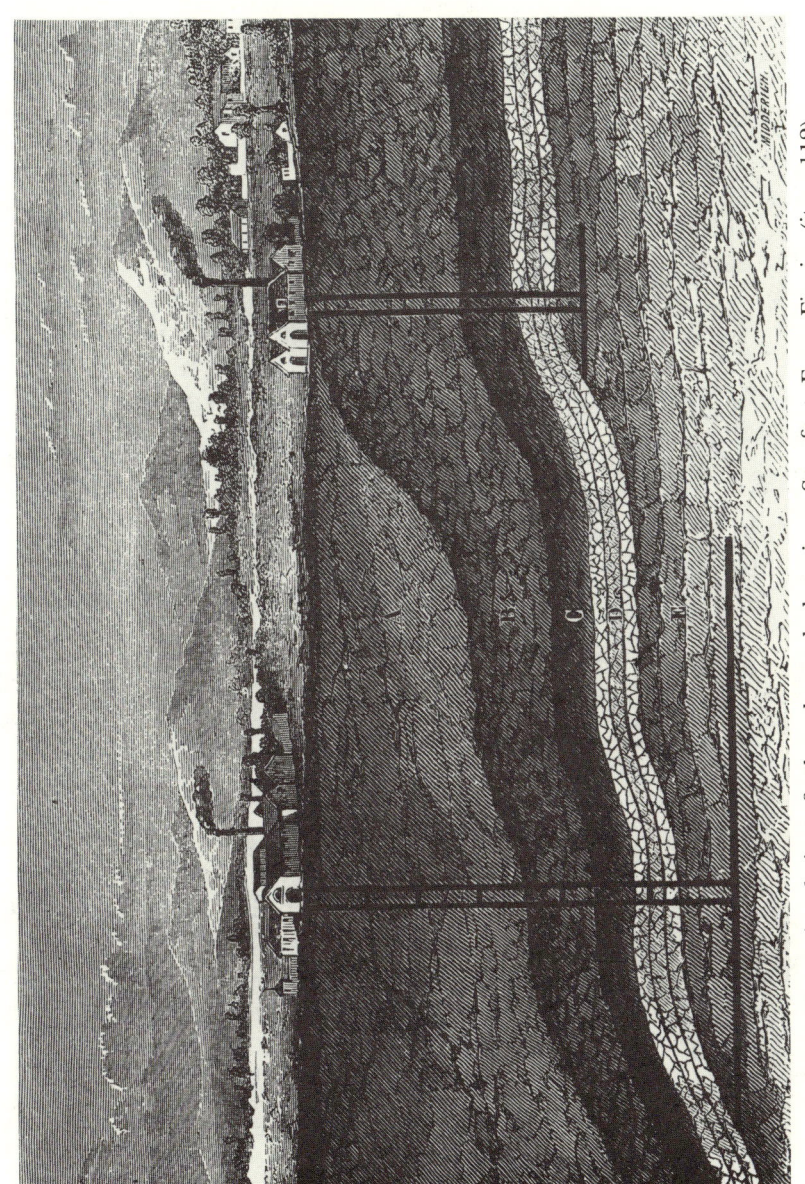

Figure 10. Sectional view of salt and potash deposits at Stassfurt. From Figuier (item 112)

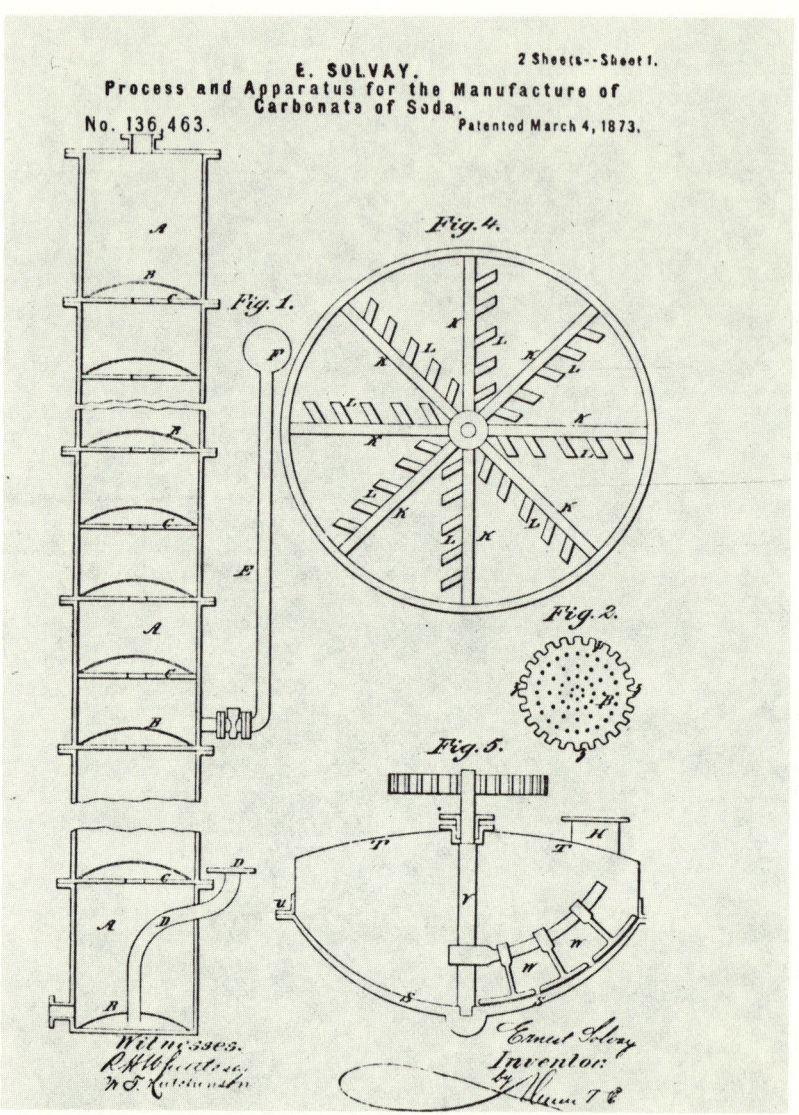

Figure 11. Solvay process for soda, as shown on the United States patent of 1873.

1453. WOOD, R. Derek. "The Daguerre Patent, the British Government, and the Royal Society," History of Photography, 4 (1980), 53-59.
Refers to peculiarities in Daguerre's English patent and the cogitations of a committee of the Royal Society on the question of awarding him a medal. Citations.

POTASH

for its history prior to about 1850 see ALKALIS (Pt. 2)

HISTORIES

1454. "Stassfurt's chemische Industrie," Berg- und Hüttmanische Zeitung, 22 (1868), 427.
A description, necessarily historical, of the mineral potash industry near Staffurt during its first decade. Probably by F. Michaels.

1455. AMERICAN POTASH PRODUCERS. Petition to the Congress of the United States Concerning the Wrecking of the American Potash Industry by the Government of Germany. New York, 1922. (43 pp.)
Not quite unbiased.

1456. BENNER, L. Wilhelm. Entwicklung, finanzieller Aufbau, und Finanzierungsmethoden der deutsche Kali-Industrie. Giessen, 1930. (115 pp.)
A dissertation with citations.

1457. BRUEHLING, Erwin. Die Entwicklung der Hannoverischen Kaliindustrie. Hanover, 1925.
A dissertation with citations.

1458. FRANK, Adolph. "Stassfurter Kali Industrie," in Hofmann, 1875 (no. 73), pp. 351-71.
Authoritative historical summary, by a "pioneer" who is said to have made the first successful boring at Stassfurt in 1851.

1459. FRANK, Adolph. "Anfang und Entwicklung des Kali bergbaues und der Kaliindustrie," Verhandlung des Vereins zur Beforderung des Gewerbfleisses, 81 (1902), pp. 233-58.

A summary, which reprints his report of 1860 on the advisability of establishing a chemical works at Stassfurt. Includes a description of its present condition.

1460. FRIEDENSBURG, F. "Kalivorkommen ausserhalb des deutschen Reichs," Kali, 6 (1912), 593-98.
A survey of recent attempts to produce potash from kelp, notably in Britain and the United States, and to find mineral potash (California and Utah). Concludes that there is no significant source outside of Germany.

1461. KRISCHE, P. "Die Geschichte der wichtigsten Kaliverbindungen," Kali, 3 (1909), 5-10.
Primarily an analysis of the most important salts in the German potash deposits.

1462. KRISCHE, P. "Fünfzig Jahre deutscher Kaliindustrie," Kali, 5 (1911), 141-46, 185-94.
A history of the exploitation of German mineral potash from the first drilling for salt (Kochsalz) at Stassfurt in 1851 to about 1865.

1463. KRISCHE, P. "Die Stassfurter Zeit des Kalisyndikats," Kali, 5 (1911), 21.
Dates the first potash mine 1857 and attributes the formation of the famous cartel to overproduction in 1864 and 1872. Describes its frequent reorganization from "The Carnallite Convention" of 1879, revealing the role of the uncertainty as to which mineral was which, in frustrating any definitive establishment of the cartel.

1464. MANSFELD, G.R., and W.B. LANG. "The Texas-New Mexico Potash Deposits," University of Texas-Bulletin, 3401, (1935), 641-832.
Includes a history of the finally sucessful search for an American mineral potash. German legislation of 1910 is said to have activated (or reactivated) the American search. Using as clues potash minerals reported by drillers for petroleum, the U.S. Bureau of Mines was mandated by Congress in 1926 to make test drillings. The results led to the establishment of production in Eddy County, New Mexico, in 1931.

1465. MITCHELL, G.E. "The Potash Search in America," <u>Review of Reviews</u>, 45 (1912), 73-77.
A review of the situation as of that date by a member of the U.S. Geological Survey. He judged that many of the reports of discovery of potash ores were "aimed to fleece the republic" and thought kelp the most promising source.

1466. PRECHT, H. "Die Entwicklung der Kaliindustrie," <u>ZAC</u>, 19 (1906), 3-7.
A sketch but notable for a list and location map of firms. No references.

1467. SCHOENEMANN, Josef. <u>Die deutsche Kali-industrie und das Kaligesetz. Ein Volkswirtschftliche Studie.</u> Hanover, 1911. (152 pp.)
A study of the history of the industry and its legal underpinnings as a guide to its improvement from the political and social viewpoints.

1468. STOCKING, G.W. <u>The Potash Industry, A Study in State Control.</u> New York, 1931. (343 pp.)
Of the German potash industry. He blames its defective postwar state on the system of state control not on the war.

1469. STOEPPEL, K.T. <u>Die deutsche Kaliindustrie und das Kalisyndikat.</u> Halle, 1904. (329 pp.)
Pp. 14-64 relate the history of the industry, from 1839 to about 1900. Citations.

1470. STUTZER-KOENIGSBERG, -. "Amerikanisches Kalisalz," <u>Kali</u>, 6 (1912), 294-95.
Reports in detail on the "energetic search" for American potash, which "has been going on for several years." Concludes that a discovery is possible but not probable.

1471. TEEPLE, John E. <u>The Industrial Development of Searles Lake Brines.</u> New York, 1929. (182 pp.)
Includes a very lively historical account of attempts to extract potash and other chemicals (notably borax) from this California desert lake from 1873. Teeple worked at Searles Lake from 1919 and gives an entertaining account of his experience. His sense of

humour is unique, in my experience, in writings on chemical technology.

SODA, ARTIFICIAL

EARLY WORKS

1472. FORDYCE, Alexander. "Patent (Aug. 1, 1781)...Process by Which the Alkali Contained in Sea Salt, Saltwater, Rock Salt, Salt Springs, Salt Cake, Glauber's Salt, and Vitriolate Tartar, is Separated from the Marine and Vitriolic Acids," Repertory of Arts and Manufactures, 4 (1796), 313-15.
This celebrated patent was granted after hearings, which were printed in the Journal of the House of Commons for May 22-June 27, 1780. There we see Fordyce producing numerous testimonials, and objections from rival claiments for prior invention, including James Keir, P.T. de Bruges, James Watt (and Joseph Black), John Collison, Joseph Fry, A. Shannon, and Isaac Cookson. The hearings concerned remission of the salt duties, to which all wanted exception, except Shannon, who claimed to be producing (artificial) "barilla" at a profit, despite the salt tax.

1473. WIEGLEB, J.C. Chemische Versuch über die alkalischen Salze. Berlin und Stettin, 2nd ed., 1781.
Begins with a 55-page history of the views of chemists on the "fixed alkali salts," after which he relates his own experiments on obtaining them from vegetable materials, on preparing them "from the supposed constituents," and on other questions.

1474. LAVOISIER, Antoine. "Prix a décerner pour l'alcalisation du sel marine," n.d., (1781 ?) In his Oeuvres (Paris, 1862-93), 6: 16-20.
Describes an "extraordinary prize" offered by the Paris Academy of Sciences for 1783. The competition was extended to 1786 and then to 1788.

1475. ACHARD, F.C. "Versuch über des Kochsalz, die vorzüglich in der Absicht es zu zersetzen und das mineralischen Alkali zu erhalten angestellt worden," In his Sammlung physiche und chemische Abhandlungen, Berlin, 1784: 93-131.

Discusses attempts which have been made to extract soda from salt including his own experiments. He concludes that it is economically impractical.

1476. WESTRUMB, J.F. "Kurze Geschichte der Scheidung des mineralischen Laugensalzes aus seinen Mittelsalzen, nebst einer Beschreibung der wohlfeilsten Bereitungsart dieses Salzes," In his Kleine Physikalisch-chemische Abhandlungen, Leipzig, 1785,1,1: 133-46.
A careful and documented description of attempts to that date to make artificial soda.

1477. HAHNEMANN, D. "Ueber die Schwerigkeit der Minerallaugensalz Bereitung durch Potasche und Kochsalz," (Crell's) Chemische Annalen, 2 (1787), 387-96.
After describing briefly other processes, he favors "the newest," the interaction of salt and potash. Unfortunately, however, eleven parts of the more expensive potash yields only seven parts of soda.

1478. SOCQUET, J.M. "Experiments et resultants de plusiers operations en grandes fait a Venise sur la preparation de la soude du sel marine," Oposcali Scelti Sulla Scienze e sulle Arti (Milan), 15 (1792 ?).
Not seen.

1479. DARCET, Jean, A. GIROUD, C.H. LELIEVRE et Betrand PELLETIER. Description des divers procédés pour extraire la soude du sel marin. Paris, 1794.
The report of the commissioners appointed by the Paris Academy to judge essays submitted for the artificial soda prize. It is summarized, with comments by other chemists, in Journal de Physique, 45 (1794), 118-34, 191-9; Bulletin de la Societé Philomathique, 1 (1794), 77; AC, 19 (1797), 58-156.

1480. SHEE, Henri. "Notes sur l'enterprise de manufacture de soude en France, par la décomposition du sel marin," Journal des Arts et Manufactures, 3 (1797), 284-99.
An appeal to the Commissioners (Darcet el al.) for support of Leblanc's soda works at Franciade (near Paris) by one of his partners.

1481. LEBLANC, Nicolas. "Mémoires et rapports concernant la fabrication du sel ammoniac et de la soude," Journal de Physique, 50 (1800), 462-71.
Two memoires by Leblanc, and comments on them by Fourcroy and Vauquelin.

HISTORIES

1482. BARKER, T.C., R. DICKINSON, and D.W.F. HARDIE. "The Origins of the Synthetic Alkali Industry in Britain," Economica, n.s. 23 (1956), 158-71.
Claims that the success of artificial alkali manufacture depended upon the maker's ability to sell the product to soapmakers.

1483. BAUD, Paul. "La manufacture de soude de Nicolas Leblanc," CR, 197 (1933), 701-03.
Brief description of the Franciade works with plan drawings.

1484. BAUD, Paul. "Les premiers soudrières Francaise," CR, 197 (1933), 1498-1500.
A succinct argument that Leblanc's rivals did not precede him in the establishment of an artificial-soda works.

1485. BRADBURN, J.A. "Der Ammoniaksodafabrikation und der Sodahandel den Vereinigten Staaten," ZAC, (1898), 14-15, 55-59, 78-83, 102-06, 149-50.
A detailed technical description without references.

1486. CHANCE, Henry. "On the Manufacture of Alkali and Acids in Birmingham and the Neighborhood," in S. Timmins, ed. The Resources, Products, and Industrial History of Birmingham, London, 1866: pp. 166-67.
Mentions three works, with production data. Says that the processes are essentially the same as they were 50 years earlier, although thanks to Gossage and the Alkali Acts the hydrochloric acid vapors are now condensed.

1487. CLAPHAM, R.C. "An account of the Commencement of Soda Manufacture on the Tyne," Trans., Newcastle Chemical Society, 1 (1868-71) 29-45.

A detailed circumstantial account including sulphuric acid manufacture. No references.

1488. COQUEBERT, Charles, "Histoire de la decomposition du sel marin," Journal des Mines, 1,3 (1795), 29-90.
Probably the earliest history of soda, this deals with both natural and artificial sources.

1489. DIZE, M.J.J. "Mémoire historique sur la décomposition du sel marin," Journal de Physique, 70 (1810), 291-300.
An account of Leblanc's work by one of his associates.

1490. DUMAS, J.B., et al. "Rapport relatif a la découverte de la soude artificielle," CR, 42 (1856), 553-78.
Report of a committee organized by the Paris Academy of Sciences to consider an appeal by the descendents of Leblanc for reparation for the damage suffered through the sequestration of his soda works (in 1794). Its conclusions favor Leblanc and explain why.

1491. EPHRAIM, J. "Die Vorlaufer von Nicolas Leblanc," AGNT, 8 (1918), 222-24.
On the question of whether earlier British patents should deprive Leblanc of credit. Ephraim concludes that they should not.

1492. GILLISPIE, Charles C. "The Discovery of the Leblanc Process," Isis, 48 (1957), 152-70.
Considered as an example of the influence of science on industrialization, he finds this case weak. The Leblanc process did not flourish because it was understood; it belongs to the history of industry not of science. Citations.

1493. GITTINS, L. "The Manufacture of Alkali in Britain, 1779-1789," AS, 22 (1966), 175-90.
Important for its analysis of the influence of the users of alkali on the introduction of artificial soda. Points out that the Society of Arts (London) offered a soda prize in 1783, the year of the famous prize offer of the Paris Academy.

1494. GOLDSTEIN, I. Deutschlands Sodaindustrie in Vergangenheit und Gegenwart. Stuttgart, 1896. (108 pp. Münchener Volkwirtschaftliches Studien no. 13)

Says that the first Leblanc soda works in Germany was that of Hermann & Sohn, Schönebeck, founded in 1840. Important for data on the industry, in England as well as Germany.

1495. GOSSAGE, W. A History of the Soda Manufacture. Liverpool, 1861 (23 pp.)
Miscellaneous information, some perhaps unique, presented to the BAAS by a pioneer of the British soda industry.

1496. HASENCLEVER, Robert, "Die Entwicklung der Sodafabrication und der damit in Zusammenhang stehenden Industrie-zweige in den letzen 25 Jahren," Berichte DCS, 29 (1896), 2861-77.
Important technical history. Citations.

1497. KINGZETT, C.T. The History, Products, and Processes of the Alkali Trade. London, 1877.
Important for cost and production data over an extended period.

1498. LUCION, R. "Einige Beiträge zur Geschichte des Ammoniac-soda-Processes," Chemiker Zeitung 13 (1889), 627.
Discusses researches of the French physicist, L. Fresnel, 1811-12, from his correspondence.

1499. LUNGE, Georg. "Geschichte der Fabrikation von weisser kaustischer Soda in Deutschland," ZAC (1896), 709-11.
Answers a reviewer of his Handbuch der Sodaindustrie (2nd ed.) who claimed that Germany produced caustic soda earlier than it was made in England.

1500. MACTEAR, James. "On the Growth of the Alkali and Bleaching-powder Manufacture of the Glasgow District," Chemical News, 35 (1877), 4-5, 14-17, 23-26, 35-36.
A circumstantial account with much data concentrating on the materials involved, salt, sulphuric acid, bleaching powder, soda, caustic soda, etc.

1501. MATTHEWS, M.H. "The Development of the Synthetic Alkali Industry in Great Britain in 1823," AS, 33 (1976), 371-82.
Considering fourteen adjacent alkali works, he dis-

cusses the factors of raw materials and markets as well as the influence of chemical education.

1502. MOILLIET, J.L. "Keir's Caustic Soda Process: An Attempted Reconstruction," Chemistry and Industry, (1966), 405-08.
The process involved the passing of a weak solution of sodium or potassium carbonate through a thick body of lime. The author's tests, with varying concentrations, leads to an estimate of contemporary economic viability. Keir's factory (ca. 1780) was at Tipton.

1503. PAYEN, (J.B.P.), "La soude artificielle," Revue des Deux Mondes, 63 (1866), 958-83.
Brief history and general discussion without citations. The author claims to have been the first, after the inventor, to have exploited the Leblanc process.

1504. PENNOCK, John D. "Progress of the Soda Industry in the United States since 1900," Berichte, V. Internationale Kongress für angewandte Chemie (Berlin), 1903, 1:661-71.
A useful miscellany of data on both artificial (Solvay) and natural soda production.

1505. SCHEURER-KESTNER, A. "Nicolas Leblanc et la soude artificielle," Revue Scientifique, ser. 3, 9 (1885), 386-96.
Detailed and circumstantial, a biographical approach using letters and unpublished papers--without saying where they were found.

1506. TRUMP, E.N. "Looking back at 60 Years in Amonia Soda Alkali," Chemical and Metallurigical Engineering, 40 (1933), 126-29.
A circumstantial history of the Solvay process in the United States by one who entered the industry at the outset (Syracuse, 1882). No references.

1507. WELDON, Walter, "On the Present Conditions of the Soda Industry," Journal, Society of Chemical Industry (London), 2 (1883), 2-12.
Not deliberately historical but an interesting description of the state of the art two decades after the introduction of the Solvay process. No citations.

SULPHURIC ACID

HISTORIES

1508. "Notice historique sur la fabrication de l'acide sulfurique," Genie Industriel, 37 (1869), 184-88.
On Michel Perret's patent (Feb. 2, 1836) for a method of burning pyrites to produce sulphur for sulphuric acid.

1509. BAUD, Paul. "John Holker et la fabrication de l'acide sulphurique," CR, 196 (1933), 1797-1800.
Discusses Holker's construction of the first lead chamber works in France, which the author dates 1769.

1510. DICKINSON, H.W. "History of Vitriol Manufacture in England," TNS, 18 (1938), 43-60.
Like no. 633, this is mistitled, as it is concerned with the establishment of the chamber process for sulphuric acid in England. Citations.

1511. DROESSER, Ellinor. Die technische Entwicklung der Schwefelsäurefabrikation und ihre volkswirtschaftliche Bedeutung. Leipzig, 1908. (220 pp.)
A history of chemical technology "only insofar as it has had economic consequences," and these seem to have only become apparent in the late nineteenth century. Still, the early history gets attention, and this is as complete a history of the sulphuric acid industry as we have.

1512. GUTTMAN, Oscar, "The Early Manufacture of Sulphuric and Nitric Acids," Journal of the Society of Chemical Industry (London), 20 (1901), 5-8.
Chiefly devoted to the production methods described in "an old manuscript book" written between 1771 and 1790 at Birmingham by W.E. Sheffield (the manuscript is now in the Guttman collection at the Hagley Library, Wilmington, Delaware).

1513. HOECHST (firm). Schwefelsäure Hoechst. Von Kammerverfahren zum Kontaktprozess, 1880-1914. Frankfurt/M, 1975. (133 pp. DHA no. 47)
Reproduces patents, drawings, and other documents with a brief introduction.

1514. LECLAIR, E. La fabrication des acides forts à Lille

Figure 12. Cross-section of chamber sulphuric acid works, with (left) a Gay-Lussac tower. From Figuier (item 112)

avant 1790. Poitiers, 1901. (15 pp.)
Not seen.

1515. LUNGE, George, "Ueber die gegenwärtige Stand der Schwefelsäureindustrie," ZAC (1903), 689-95.
An authoritative description of the state of the field.

1516. MACTEAR, James, "Address (on the history of sulphuric acid manufacture)," Proc., Philosophical Society of Glasgow, 13 (1882), 409-27.
A chatty account, with much undocumented quotation, on sulphuric acid manufacture to about 1840. Lists works in Britain.

1517. MOELLER, Wilhelm. Geschichte der Entwicklung der K.K. Schwefelsäure-Fabrik in Unter-Heiligenstadt.
Vienna, 1895. (14 pp. Monographien des Museum für Geschichte der österreichischen Arbeit, Hft. 6)
Detailed and illustrated description, without citations.

1518. RAUTER, Gustav, "Der gegenwärtige Stand der Schefelsäure-industrie," SC, 8 (1903), 257-302.
On the state of the field, with numerous citations, especially to patents.

1519. VOGEL, O. "Auf Geschichte der deutschen Schefelsäureindustrie," Chemiker Zeitung, 58 (1934), 472.
A note referring to the production of oil of vitriol in Silesia in the eighteenth century.

1520. WAESER, Bruno. Schwefelsäure/Sulfat/Salzsäure. Dresden & Leipzig, 1927. (121 pp. Fortschritte in der Chemische Technologie in Einzeldarstellung, Bd. 11)
Intended to save the reader the trouble of paging through long series of annual reports (Jahresberichten). History is a secondary aim but a necessary consequence. Numerous citations.

1521. WAESER, Bruno. Handbuch der Schwefelsäurefabrikation.
Braunschweig, 1930. 3 vols.
A revision of the famous handbook of George Lunge. Vol. 1 gives some history.

1522. WEDGE, Utley. "The Sulphuric Acid Industry in the United States," Proceedings, 8th International Congress of Applied Chemistry (Washington-New York,

1912), 2: 241-48.
Summary of the state of the art, with particular reference to its uses.

1523. WELLS, Arthur E., and D.W. FOGG. The Manufacture of Sulphuric Acid in the United States. Washington: GPO, 1920. (216 pp.)
Gives production data from 1865 to 1918. Not otherwise historical.

1524. WINKLER, Clemens. "Die Entwicklung der Schwefelsäure-fabrikation im Laufe des scheidenden Jahrhundert," ZAC, (1900), 731-39.
A sketch, beginning with W.A. Lampadius' Grundriss der technischen Chemie (1815) and ending with Winkler's own work on the contact process in the 1870's. No citations.

TEXTILE CHEMISTRY

HISTORIES

1525. MARSH, J.T. Mercerising. London, 1941. (474 pp.)
History, pp. 1-27. John Mercer (d. 1866) introduced this method of finishing cotton cloth by treatment with caustic soda.

1526. PARK, G.H., and E. GLOUBERMAN. "The Importance of Chemical Developments in the Textile Industries during the Industrial Revolution," JCE, 9 (1932), 1142-70.
Deals with bleaching and dyeing, mainly in Britain in the eighteenth century.

1527. ROBINSON, Stuart. A History of Dyed Textiles. London: Studio Vista, 1969. (112 pp.)
A straightforward topical-chronological history with citations, illustrations (color), and a list of museums exhibiting textiles.

1528. TUTTLE, F.J. "The Story of Coated Fabrics," Textile Research, 14 (1944), 228-32, 260-69.
The first part deals with the development of "oil cloth," a name which appeared in 1694, long after the first relevant (English) patent (1636). The second part deals with rubber and pyroxylin coatings from the work of Charles Macintosh (1823). References.

AUTHOR INDEX

Abrahams, H., 350
Accum, F., 1298
Achard, F.C., 1475
Acheson, E.G., 1113, 1263
Adams, J.Q., 236
Adrosko, R. J., 725
Agricola, G., 1
Aigner, A., 989
Alberti, M., 643
Alberti, W.C., 963
Allen, Hugh, 482
Allen, H. W., 776
Allen, J. A., 1162
Allen, J.F., 396
Althin, T., 515
American Chemical Society, 351
American Chemical Society, Div. of Petroleum, 1361
American Gas Association, 1299
American Potash Producers, 1455
Anastasi, A., 433
Ancelon, E.A., 990
Anderbjork, J.E., 499
Anderson, J. 887-88
Anderson, J.D., 596
Anderson, D.E., 777-78
Andre, C.C., 191
Andre-Félix, A., 316
Andréossy, A.F., 594, 1066
Andrew, E. E., 991
Appert, N., 773
Aremarius, J.C., 1163
Armstrong, H.E., 437
Armstrong, W. G. 277
Arnold, J.P., 779
Arntz, H., 780
Ash, C.S., 352
Ashton, F., 992
Aslund, B., 1072
Astarita, G., 92, 320
Atkinson, R.W., 155
Atwood, A.W., 472
Ayerst, R.P., 1273
Aynsley, E.E. 558

B.D., 1199
Baar, L., 238
Babcock, G.D., 530
Bacon, J.B.F., 827
Bader, L., 1300
Baeyer, A., 1221
Baier, W. 1431
Baldry, P.E., 1384
Baldwin, R.T., 1320
Balland, A., 1073
Baltz, C., 993
Bancroft, E., 726
Barker, D. H.. 156
Barker, P. W., 947
Barker, T.C., 278, 514, 1482
Baron, S.W., 781
Barth von Wehrenalp, E., 1164
BASF, firm, 93, 453, 877
Bathe, G., 994
Baud, P., 214, 215, 216, 1483, 1484, 1509
Bauer, A., 74, 174
Bauer, R., 1165, 1166
Baumhauer, E.H., 315
Baumler, E., 489
Baxa, J. 1100-01
Baxter, J.P. 3rd, 353
Bayer, firm, 461
Beaton, K., 522
Bebbington, P.S., 423
Beck, R., 632
Beckmann, J., 9, 19, 29, 94
Bedhome, R., 217
Beer, E.J., 1167
Beer, J.J., 727, 1222-23
Bell, J., 60, 1385
Benaerts, P., 239
Benedik, T.G., 1413
Benjamin, M., 354
Benner, L.W., 1456
Benton, W.E., 672
Bergengren, E., 442
Bergman, T.O., 617
Berl, E., 436
Bermudez Miral, O., 1074
Bernsthen, A., 1321, 1351

Bertelsmann, W., 1301
Berthelot, M., 95
Berthollet, C. L., 714, 1201, 1205
Beyer, K.G., 852
Biasutti, G.S., 559
Bigelow, J., 50
Binz, A., 96, 240, 903
Bird, R.B., 97
Birdseye, C., 782
Biringuccio, V., 1
Birkenfeld, W., 1212
Bischof, F., 995
Bischof, J.N., 728
Bishop, J.L., 355
Bitting, A.W., 783
Bladen, V.W., 673
Blair, D., 157
Blakey, A.F., 1120
Bloch, M.R., 996-97
Blockmann, G.M.S., 1302
Bloede, V.G., 351
Bodman, G., 597
Boeheim, W., 850
Boerhaave, H., 5
Böttger, F., 678
Bogdan, I., 914
Bollaert, W., 653
Bolle, J., 524
Bollen, P.A., 1224
Bolton, E.K., 1168
Bolton, J., 428
Bondois, P.M., 828
Borelli, ---, 574
Born, W., 729
Borscheid, P., 241
Bosch, C, 1145
Bottée de Toulmont, J.J.A., 1069
Boussingault, J.B.J.D., 1317
Bowles, W., 337
Bowring, J., 649
Boyle, R., 940
Brace, H.W., 904
Braconnet, H., 1289
Bradburn, H., 1486
Bradford, C.W., 951

Brand, C.J., 1121
Brandenburger, K., 1169
Bravo, G.A., 878
Brenemann, A.A., 1322
Bretschneider, E., 158
Bridbury, A.R., 998
Briggs, H., 1216
Bright, R., 192
Brightman, R., 1225
Brock, A. St. U., 933
Brown, H.T., 784
Brown, J., 703
Brown, S., 1068
Browne, C.A., 98, 356, 405, 598, 1102, 1122
Brownlie, D., 1303
Brownrigg, W., 981
Bruchling, E., 1457
Brunck, H., 1226
Brunello, F., 730
Buchner, M., 99
Bucholz, K., 100
Buck, A.J., 1170
Budapest, Technical University, 195
Bürgin, A., 479
Bugge, G., 397
Bujard, A., 1274
Bullier, L.M., 1345
Burstyn, H.E., 1123
Bussolin, D., 829
Bustelo Vazquez, F., 340

Caley, E.R., 731
Callahan, J.R., 373
Calvert, A.F., 999
Cameron, F.T., 416
Campbell, A., 588
Campbell, M., 417
Campbell, M.R., 655
Campbell, W.A., 101, 279, 558
Cancrin, F.L., 986
Canepario, P.M., 611
Capocaccia, A., 102
Cardwell, D.L.S., 754
Carle, W., 1000
Caro, H., 242, 426, 1227

Carpenter, C., 357
Cartwright, A.P., 451
Cato, M.P., 882
Cavillier, ---, 627
Cayley, E.R., 830
Cederhielen, C.W., 883
Chagnon, A., 218
Chambon R., 831
Chance, H., 1486
Chance, J.F., 464
Chance, W.H.S., 832
Chandler, D., 1304
Chang Siu-Ming, 159
Chaptal, J.A.C., 41, 208, 211, 218, 540, 544, 567, 717, 732
Charleston, R.J., 833
Chassagne, S., 1525
<u>Chemical Age of India</u>, 160
<u>Chemical Society (London)</u>, 398
Chevallier, M., 69
Chilton, T.H., 1352
Choffel, J., 517
Christiansen, C.C., 103, 243
Christofle, C., 1264
Christophe, R., 1146
CIBA, 343-44, 466
Cibot, P.M., 647
Clapham, R.C., 1487
Claquesin, P., 785
Clark, H.O., 733
Clark, J.A., 1370
Clark, J.D., 1293
Clark, V.S., 358
Clarke, M.J., 1353
Class, W., 915
Claude, G., 1323
Clegg, S., Jr., 1305, 1311
Clements, R., 104
Cliffe, W.H., 1228-29
Clow, A., 280
Clow, A. & N., 105, 421, 445, 599, 633
Cnoll, S.B., 643
Cochin, A., 834
Coe, B., 1432
Cohen, J.M., 438
Cohn, P., 91

Coker, W.S., 560
Colby, L.J.M., 734
Cole, A.H., 14
Collet-Descostils, H.V., 970
Collie, M.F., 1114
Colson, A., 106
Colwall, D., 614
Comroe, J.H., 1386
Condamine, C.M., de la, 943
Coppock, J.B.M., 786
Coquebert, C., 1488
Cordier, L., 179
Cornwall-Clyne, A., 1433
Corran, H.S., 787
Coste, P., 218
Cracroft, P.G., 1306
Cramer, H., 1001
Crane, J. E., 359
Crass, M.F., Jr., 1275
Crawford, E.T., Jr., 360
Crookes, W., 58
Crossley, ---, 1230
Crutznach, L.S., 886
Cunningham, A., 651
Cutting, C.L., 788

Däbritz, W., 481, 503
Daquerre, L.J. M., 1430
Dahl, C.F., 916
D'Alembert, J.D., 8
Darcet, J.P.J., 542, 1479
Daumas, M., 107
Davis, A.C., 890
Davis, P., 835
Davy, H., 735, 1117
Davy, R., 219
Dawson, T.R., 955
Deblinne, ---, 889
Deerr, N., 1103-04
De Haen, P., 1387
De La Beche, H.T., 551
Delumeau, J., 634
Demachy, J.F., 18
<u>Descriptions des Arts et Métiers</u>, 13
DeSousa, J.P., 161

Dickinson, H.W., 905, 1510
Dickinson, R., 1482
<u>Dictionnaire Technologique</u>, 47
Diderot, D., 8
Dick, W.F.L., 462
Diem, A., 133
Dietz, D., 483
Disanti, E., 789
Divers, E., 162
Dixon, W.H., 1276
Dize, M.J.J., 1489
Djermantovich, R., 108
Dluqosz, A., 1003
Dobbin, L., 1324
Dobson, E., 671
Dodd, G., 275
Donald, M.B., 1075
Donath, E., 196
Donop, ---, 889
Dorenfeldt, L.J., 916
Dossie, R., 583
Douglas, R.W., 836
Downard, W.L., 790
Drefus, H., 1171
Drôsser, E., 1511
Dubois, J.H., 1172
Dubpernell, G, 1265
Dubrenfaut, A.P., 1148
Dufay, (C.F?), 577
Duhamel, J.P.F., 235
Duhamel du Monceau, H.L., 578, 982
Duisberg, C., 418, 419, 1231
Dujardin, J., 1149
Dumas, J.B., 49, 1490
Dumont, G., 1325
Duncan, G.S., 837
Dundonald, 9th Earl of, 274, 985, 1214-15
Dunlap, J.R., 1124
Dupont, firm, 1173
Dupree, A.H., 361
Dupuqet, ---, 625
Dutton, W.S., 473, 1277
Dyer, B.W., 656

Eason, A., 1200

Eastman, W., 906
Edelen von Kees, S., 193
Edelstein, S., 399, 736
Eder, J.M., 1434
Eibner, A., 737
Eichgruen, A. 1388
Eisen, G.A., 838
Elder, A.L., 1389
Elektrochemisk, firm, 476
Elton, A., 1307
Emerson, E.R., 791
<u>Encyclopedie Methodique</u>, 25
Ender, W., 1254
Engelhorn, F., 422
Engle, A., 839
Engstrom, E., 620
Enhalbert, H., 792
Enos, J.L., 1371
Ephraim, J., 1491
Eriksen, S., 674
Ernst, E., 1390
Erxleben, J.C.P., 228, 621
Evans, K.A., 1002
Evelyn, J., 534
Everard, S., 477
Eversmann, F.A.A., 311
Eyer, F., 793
Eyre, J.D., 163
Ezell, J.S., 498

Faber, G.A., 738
Fabricius, J.C., 272
Faehler, E., 934
Farber, E., 128, 399
Faugas de Saint Fond, 273
Fauque, M., 220
Fauquet, L.G., 1174
Faust, B., 1076
Favé, E., 941
Fayol, A., 434
Fechner, H., 244
Feiler, P., 1217
Feldhaus, F.M., 1150
Feldman, H.A., 1391
Felix, D.A., 917

Ferber, J.J., 109, 309, 319
Ferguson, J., 110
Fester, G., 111
Fichtel, J.E., 1004
Fieser, L.F., 1232
Figuier, L., 112
Filby, F.A., 794
Fischer, Emil, 426
Fischer, Ernst, 400
Fischer, H., 1175, 1278
Fischer, W., 33, 532
Fitzgerald, G.A., 728
Fiumi, E., 657
Flechtner, H.J., 420
Fleming, H.W., 409
Flick, C., 561
Florentiis, G., 113
Flurl, M., 1005
Foderé, F.E., 543
Foerster, J.C, 1006
Fogg, D.W., 1523
Forberger, R., 245
Forbes, R.J., 114, 521, 795, 1151, 1373
Forbin, V., 949
Fordemwalt, F., 1233
Fordyce, A., 1472
Forgeroux, ---, 538
Forrestal, D.J., 508
Forsing, P., 840
Forstmann, W., 486
Foster, J., 675
Fourcroy, A.F., 23
Fourcroy de Ramacourt, C.R., 885
Fourdriner, S., 386
France, govt., 546, 1062
France, Comite de Salut Publique, 1065
Francis, A.J., 891
Francis, C., 796
Franck, H., 1355
Frank, A., 658, 1125, 1458, 1459
Frank, S., 836
Freeman, C., 1176
Frémy, E., 841
Freshwater, D.C., 115
Fresnius, C.R., 720

Frey, C.N., 797
Freydank, H., 1007-08
Frick, ---, 549
Friedel, R., 1177, 1178
Friedensburg, F., 1460
Friedman, J. S., 1435
Fries, A.A., 1326
Fuchs, J.L., 646
Fuhrmann, J.T., 331
Fulda, E., 1010
Fulmer, J., 281
Fursen, O., 1009
Furter, W.F., 116

Gädicke, J.C., 263
Gale, E., 164
Gamble, J.C., 396
Ganzenmuller, W., 117
Garbett, S., 423
Gardner, W.H., 1179
Gardner, W.M., 1234
Garrod, L P., 1392
Gasparetto, A., 842
Gatterer, C.W.J., 231
Gedickens, J.C., 967
Geer, W.C., 950-51
Gehring, P., 1010
Geigy, firm, 479
General Chemical, Co., 480
Geoffroy, C.J., 961
Geoffroy, E.F., 704
Geoffroy, E.G., 615
George, S.L., 1279
Gerhardt, D., 1234
Gernsheim, H., 1436
Gerschel, L., 739
Gerstley, J.M., 659
Gesellschaft für die Geschichte und Bibliographie des
    Brauwesens,798
Gessner, A., 892
Gettens, R.J., 740
Gibbons, W.A., 952
Gibbs, F.W., 282, 741, 1095
Gibson, R.O., 1180
Gicklhorn, J., 1077

Giddins, P.H., 1372
Giebelhaus, A.W., 1294
Giedion, S., 799
Giffei, K.A., 1078
Gilbert, J.H., 1126, 1127
Gildenmeister, E., 907
Gillins, L., 600
Gillispie, C.C., 221, 1492
Ginori - Conti, P., 321
Giorgi, J.G., 327
Girard, C., 78
Gittins, L., 1493
Glaser, C., 413
Glaser, J., 197
Glassford, C.F.O., 601
Glauber, 2, 227
Gleditsch, J.G., 585
Glouberman, E. 1526
Gmelin, C.G., 719
Gmelin, J.F., 26
Godfrey, E.S., 843
Goidsenhoven, J.P., 165
Goldblith, S.A., 800
Goldschmidt, G., 76
Goldstein, I., 1494
Goldwater, L.J., 562
Goodfield, J., 1152
Goodrich, L.C., 166
Goodyear, C., 944
Gordy, L., 365
Gornowski, E.J., 476
Gossage, W., 1495
Gotthard, J.C., 36
Gottschalk, A., 801
Gower, H.F., 660
Graaf, P., 918
Grandeau, M.L., 67
Gravenhorst, J.H. & C.J., 708
Gray, S.F., 48
Great Britain, govt., 534-35, 802
Greenfield, J., 1189
Gregory, K., 1079
Greup-Roldanus, S., 1209
Grevel, W., 1012
Grinder, R.D., 563

Grossmann, H., 288, 362, 1354-55
Grotkass, R.E., 1105
Grünzweig, C., 403
Guedon, J-C, 118, 363
Guenther, W., 1013
Guerra, F., 1393
Guilloud, G.L., 51
Guinot, F., 119
Guttmann, O., 935, 1512
Guyton di Morveau, L.B., 23, 540, 544
Gyllenborg, G.A., 1115

Haber, F., 89, 120, 246, 1266
Haber, Z.F., 121, 122
Hadley, E.J., 452
Hagen, T.V., 622
Hahnemann, D., 1477
Hahnemann, S., 18
Halahan, B.C., 844
Halasz, N., 441
Halden, J., 936
Hale, H., 364
Hale, J.P., 1014
Hall, A.R., 1127
Hall, M.B., 283
Halle, J.S., 12
Haller, A., 82, 85
Halstead, P.E., 893
Hamilton, A., 564
Hanbury, D., 167
Hancock, J., 1061
Hangarter, W., 1394
Hanks, H.G., 661
Hard, A.H., 1181
Hardie, D.W.F., 284-87, 1327, 1482
Harding, T.C., 1128
Hare, Ronald, 1395
Harmant, P.G., 1437
Harris, J.R., 278
Hart, E., 662
Hartmann, ---, 803
Hasenclever, R., 1496
Hässenstein, W., 1280
Hassinger, H., 198
Hatch, C.E. Jr., 845

Hatton, H., 417
Haudiquer de Blancourt, J., 826
Hausbrand, ---, 742
Hausman, J.F.L., 743
Havard, F.T., 676
Hawley, L.F., 908
Hayes, S.D., 1374
Hayes, T., 539
Haynes, E.B., 846
Haynes, W., 123, 365-66, 401, 429
Heaviside, M, 1281
Hedfors, H., 695
Heers, J.L., 635
Héger, P.F., 446
Heinitz, F.A., 232
Heintz, ---, 677-78
Helbig, E., 124
Helck, W., 804
Hellot, J., 705
Helmershaufen, R., 698
Henahan, J.F., 367
Henckel, J.F., 537
Henderson, A., 805
Hendrick, E., 351
Henglein, F.A., 1328
Henkel, firm, 483
Henshaw, T., 1059
Heraclius, 697
Herbert-Kerchnawe, E., 199
Hermbstadt, S.F., 44
Hernandez Cornejo, R., 1080
Hesse, A., 288
Heuck, K., 1362
Heusser, A.H., 744
Higgins, S.H., 1210
Hildebrandt, F., 37
Hinterlang, C., 42
Hippisley, ---, 679
Hocquet, J.C., 1015
Hodgetts, E.A.B., 1282
Hoechst, firm, 125, 247, 490, 1182, 1183, 1235
    1236, 1356, 1396, 1397, 1398, 1438, 1513
Hoefer, F., 644, 664
Hoelscher, F., 1184
Hoettenroth, V., 1185

Hoffmann, F., 536
Hoffmann, G.A., 10
Hoffman, P., 484
Hofmann, A.W., 64, 68, 73, 426
Holdermann, K., 411-12
Hollander, S., 1186
Holzzch, K., 1237
Homberg, W., 575, 940, 960
Home, F., 1198
Hondorff, F., 978
Hopff, H., 1187
Hopson, C.R., 21
Horn, G., 847
Hougan, O. A. 368
Houseman, C.R., 1329
Hovestadt, H., 848
Howard, F.A., 953
Hrdina, J.N., 1016
Hubbard, W.N., 1399
Hubon, E., 1325
Hughes, E., 1017-18
Hughes, T.P., 1218
Hulme, E.W., 849
Hume, J.R., 289
Hunt, C., 1308
Hunt, R., 1439
Hunter, D., 919-20, 925
Hunter, H.L., 910
Hurry, J.B., 745
Hussey, M., 533
Huth, H., 746

ICI, firm, 1363
Idhe, A.J., 126
Ilg, A., 850
International Association of Ice Cream Manufacturers, 806
Ipatieff, V.N., 431
Issekulz, B., 1400
Ivey, D.B., 369

Jackson, W., 977
Jacob, F., 422
Jacob, H.E., 807
Jacob, S., 248
Jacquet, N., 345

James, T.H., 1444
Jamison, A., 1313
Janke, F., 1355
Jardine, R., 444
Jaubert, G.F., 346
Jeffery, T.P., 1330
Jenkins, R., 636, 680, 1440
Jenkins, T.E., 79
Jimbo, G., 168
Johns, C., 1375
Johnson, H.C.E., 370
Johnson, J.A., 249
Johnson, R.P., 747
Jones, D.P., 371
Jordi Gonzalez, R. 341
Journal of Glass Studies, 851
Juenger, W., 954
Julien, S., 169, 170
Jung-Stilling, J.H., 24
Jussieu, A., 576
Justi, J.H.G., 9, 14, 706

Kämpfer, F. 852
Kalbfleisch, M., 401
Kalkschmidt, E., 681
Kapff, F., 715
Karsten, D.L.G., 1215
Kasteleijn, P.J., 14, 27
Kaswell, E.R., 748
Kaufman, M., 1188
Kaufmann, C.B., 408
Kaufmann, G.B., 1401
Keckowa, A., 1019
Keim, K., 921
Kellner, D., 979
Kelly, F. C., 467
Kendall, E.C., 1402
Kenyon, G.H., 853
Keppeler, G., 86
Keppler, J.A., 1163
Kerr-McGee, firm, 498
Kersaint, G., 222
Keuchel, E.F., 808
King, P.E. 749
Kingzett, C.T., 1497

Kirkby, W., 809
Kirkpatrick, S.D., 372-73
Kisa, A., 854
Klaiber, H., 1020
Klaproth, D.M.F., 629
Klein, H., 1021
Kleinschrod, C.T., 1022
Klinckenstroem, C.G., 602, 1023
Klippert, ---, 1129
Klipstein, K.H., 1230
Knapp, F.L., 56
Knight, C., 59
Knowles, J.A., 855
Knox, ---, 631
Koch-Sternfeld, J.E., 1024
Kocher, E., 171
Koerner, ---, 322
Koerner, G., 1025
Koehler, A.W., 624
Koerner, T., 879
Körting, A., 1309
Kolb, J., 127, 223
Konrad, E., 1364
Kopp, U.F., 1026
Kotowski, A., 227, 937
Kränzlein, P., 492
Krammer, A., 374, 1295
Kranzberg, M., 128
Kremers, E., 1403
Krenkow, F., 682
Kreps, T.J., 603
Krisch, Paul, 1130, 1131, 1461-63
Krünitz, G.F., 38
Krünitz, J.G., 16
Kuennert, H., 520
Kuhlmann, firm, 500
Kunckel, J., 826
Kunradt, J.A., 29
Kunzmann, T., 250
Kurtz, H., 1027

Laer, J., 251
Lacoin, M., 218
Lafuente y Poyanes, M., 1106
Lamberville, C., 1211

Lampadius, W.A., 46
Lamprecht, G.F., 28
Lancisi, J.M., 318
Lang, A., 1331
Lang, W.B., 1464
Langsdorf, K.C., 887
Lardner, D., 856
Laufer, B., 172
Lauferbach, F., 750, 1238
Lauth, C., 80
Lavoisier, A., 1474
Lawrence, A.A., 1376
Lawrie, L.G., 751
Leather, J.W., 1081
Learitt, R.K., 513
Leblanc, N., 1206, 1481
Leclair, E., 604, 1514
Lefébure, C., 446
Legett, W.F., 752
Lehmann, J.G., 707
Leibniz, G.W., 940
Leif, I.P., 922
Leighton, H., 686
Leix, A., 753
Lemery, L., 616
Lemery, N., 641
Lemonnier, P., 1028
Lenoir, H., 1082
Leonhardi, J.G., 16
Lepsius, B., 252
Lequir, ---, 1107
Lerner, F., 857
Lescher, H.Z., 1239
Lesley, R.W., 894
LeSueur, E.A., 1267
Levenstein, I., 1240
Levey, M., 637, 810, 895, 1153
Lewis, H.C., 435
Lewis, W., 14, 584
Li Ch'iao P'ing, 173
Li Jung, 174
Liddell, W.A., 858
Liebenau, J. M., 1404
Liebig, Justus, 1119
Lief, A., 477

Lightfoot, J., 1241
Lippmann, E.O., 129, 130, 811-12, 909, 1108, 1154
Lischka, J.R., 290
Lister, M., 1060
Little, A.D., 131
Lombard, H.C., 550
Lord, V.A., 415
Lorio, M., 323,, 1132
Lotz, A., 938
Loysel, J.B., 1204
Lucion, R., 1498
Ludewig, W., 454
Ludwig, A., 1346-47
Lunge, G., 1268, 1499, 1515
Lyell, C., 962

McCallen, S.E., 1133
McDermott, C.H., 880
McLaughlan, G.A., 1134

Macarel, L.A., 565
MacCloskey, J.E. Jr., 487
MacDonald, G.W., 1283
Marfarlan, J., 1310
Mach, E., 131, 457
MacLeod, J.J.R., 1405
MacLeod, R.M., 566
Macquer, P.J., 7, 15
MacTear, J., 291, 1500, 1516
Mahoney, T., 375
Maierhofer, J., 133
Mallet,. J.W., 77
Mansfeld, G.R., 1464
Mantignon, C., 1347, 1348-49
Marcus, A., 449
Marion, G., 1406
Marryat, J., 683
Marsh, J.T., 1525
Marshall, A., 1284
Martel, P., 200, 253-54, 1029
Marx, C., 1189
Mathews, L.G., 1406
Mathews, W., 1311
Mathias, P., 813
Matthews, M.H., 1501

Matthis, A.R., 407
Mauries, M., 820
Maurizio, A., 814
Mayer, E., 605
Mayers, W.F., 175
Mazeas, G., 618
Mecht, O., 261
Medola, R., 1242
Mees, C.E.K., 1441
Meidinger, K. 593
Mellor, C.M., 754
Mendeleer, D. I., 332
Mercati, M., 318
Merck, E., 147
Merimee, C., 716
Merores, M., 1030
Merrifield, M.P., 699
Merritt, C., 826
Meyers, L., 1135
Miall, L.M., 1407
Miall, S., 292
Michael, T.H.G., 376
Mienes, K., 1190
Miles, W., 402
Minderer, R., 610
Minutoli, H., 755
Mitchell, G.E., 1465
Mitchell, J., 579
Mitra, C.R., 156
Mittasch, A., 1332, 1357
Modes, C.H., 859
Moeller, W., 1517
Moilliet, A., 432
Moilliet, J.L., 1502
Mollat, M., 1031
Moller Nicolaisen, N.A., 923
Mond, L., 1333
Monnet, A.G., 619
Monroe, C.A., 351
Monroe, D., 586
Monteil, J., 567
Moore, R.B., 1334
Morand, J.F.C., 1212
Moray, R., 612
Morgan, G.T., 293

Morton, J., 1243
Mosley, C.G., 1377
Mukerji, J.N., 1081
Munsell, J., 924
Mullin, B.F., 910
Multhauf, R.P., 128, 606, 663, 972, 1032, 1083
Murray, R.L., 1335
Muspratt, J.S., 55
Mussen, A.E., 294, 469
Muthesius, V., 1191
Myers, J.W., 1033

Nagel, A., 1244, 1296, 1336, 1365
Nasini, R., 664
Naturhistorische und Chemisch-Technische Notizen, 134
Needham, J., 176
Nef., J.U., 295-96
Nemnich, P.A., 210, 312
Nenquin, J., 1034
Neri, A., 826
Nettleton, L.C., 1041
Neuberg, F., 860
Newbold, B.T., 388
Newman, M., 684
Nicholls, V.V., 387
Nicholson, W., 31
Nietzki, R., 1245
Nixon, H.C., 815
Nobel, J.V., 756
Noel, S.B.J., 1042
Norris, W.G., 568
Noyes, W.A., Jr., 377
Nungesser, J.R., 758

O'Beirne, D.R., 521
Oberdorfer, K., 403
Olbrich, H., 816
Oliver, T., 569
Olsen, K.A., 511
Oppenheim, A.L., 803, 861
Orchin, M., 1192
Ostroff, E., 1442, 1443, 1444, 1445
Ostwald, W., 395
Otruba, G., 202

Pacific Coast Borax, firm, 665
Packard, W.G.T., 1136
Padley, R., 297
Pajot Descharmes, C., 545
Pallas, P.S., 326, 330
Pantyuchoff, N., 333
Papiergeschichte, 925
Papy, L., 1043
Park, G.H., 1526
Parker, A., 298
Parker, E.W., 1035
Parkes. S., 45
Parnell, E.A., 54
Partington, J.R., 135, 758, 939-40
Pasley, C.W., 896
Patel, B., 177
Patzer, H., 1408
Payen, A., 47, 57, 650, 1208
Payen, J.B.P., 1503
Payen, R.., 1084
Peebles, M.W.A., 1312
Peligot, E., 862
Pellitan, P., 541
Pelt, J.M., 1409
Penninger, E., 982, 1036
Pennock, J.D., 1504
Perkin, W.H., 430, 1246, 1247, 1248, 1249
Perlick, A F.W., 1358
Peter, C., 1037
Peters, H., 685
Peterson, E., 475
Petri, J.C., 329
Petzina, D., 496
Pfeiffer, H.H., 1213
Pfeiffer, J.F., 20
Phillips, J.A., 654
Picard, A., 81
Piccolopasso, C., 670
Pictet, R., 1337
Pier, M., 1297
Pietsch, E., 424
Pinchart, A., 863
Pinnow, H., 460, 505
Pistor, G., 485
Playfair, L., 60, 551

Ploss, E.E. 759
Poppe, G., 1038
Poppe, J.H.M., 40, 136, 548
Porter, H.L., 48
Potonniée, G., 1446
Pound, A., 507
Poutet, J.J.E., 1096
Powell, H., 864
Prange, C.F., 760
Pratt, D.D. 293
Pratt, J.D., 287
Precht, H., 1466
Priebe, P.M., 1401
Priesner, C., 1039
Prieto, Matte, J.J., 1085
Prieur, A., 1086
Prinet, M., 1040
Pris, C., 519, 865
Pritchard, D.A., 1338
Prochazka, G.A., 351
Proust, J.L., 648, 1067
Pummerer, R., 1366
Pursell, C.W., 128, 1137

Qettel, F., 1269
Queeny, J.F., 401
Queisner, R., 459
Quietmeyer, F., 897

Rabkin, V.M., 1378, 1379
Rademacher, F., 866
Rafaello, N., 666
Rahman, A., 178
Raichle, L., 1155
Ramazzini, B., 535
Ranc, A., 1087
Raoul, Y., 817
Raschen H., 484
Rastell, T., 977
Ratekin, M., 1109
Rath, U., 1410
Rauter, G., 1518
Read, J., 1251
Reader, W.J., 493
Redwood, T., 1385

Reilly, D., 137
Reinaud, J.T., 941
Reisenegger, K., 1088
Reisert, A., 1250
Remusat, A., 179
Rennewald, H.C., 642
Reutter de Rosemont, L., 1411
Reymersholms, firm, 515
Rhees, A., 34
Richard, C., 1089
Richards, A.N., 1412
Richards, J.W., 1270
Richardson, H.W., 299, 300
Rideal, C.F., 472
Riehm, K., 1044
Riemann, M., 68
Rienzi, E., 324
Ries, H., 686
Riffault des Hetres, J.R.D., 1069
Riley, J.J., 818
Rinmann, S., 711
Roberts, C.N., 687
Roberts, W.J., 607
Robinson, E., 294
Robinson, R., 443, 1251
Robinson, S., 761, 1527
Robiquet, P.J., 47
Roche, M., 342
Roderick, G.W., 439, 570
Rodnan, G.P., 1413
Roellig, W., 819
Roesling, C.L., 39, 969
Roessig, C.G., 30
Rössling, C.L., 40
Roggersdorf, W., 453
Rohland, ---, 898
Romocki, S.J., 1285
Roscoe, H.E., 1252
Rose, F., 254
Rosen von Rosenstein, N., 587
Rossiter, M.W., 1138
Rossman, J., 899
Rothe, ---. 968
Rowe, F.M., 1253
Rue, ---, 926

Ruf, S., 1045
Ruffin, E., 1118
Ruffner, D., 987
Rupp., T.L., 1202
Ruska, J., 973-74
Ruske, W., 255
Russell, E.J., 1139

Sabück, H., 440
Sacc, ---, 762
Sadtler, S.P., 378
Saechtling, H.J., 1164, 1193
Salzmann, L.F., 301
Sanderman, W., 927
Sarker, J.N., 1090
Sarneki, K., 334
Sasoly, R., 494
Savare, J., 911
Schaaf, L., 1447
Schadewalt, H., 1414, 1415
Schaefer, G., 1097
Schaefer, E., 180, 181
Schall, H., 256
Scharroo, P.W., 900
Schebek, E., 867
Scheele, C.W., 710, 712
Schelenz, H., 1416
Scheler, L., 1091
Scherer, J.A., 209
Scheurer-Kestner, A., 1505
Schicht, H., 203
Schinck, E., 302
Schlatter, H., 1290
Schleiden, M.J., 1046
Schlenk, O., 1255
Schliesser, K., 459
Schmauderer, E., 138, 139, 763, 912, 1098
Schmidt, A., 140
Schmidt-Pauli, E., 410
Schmitz, P.M.E., 1380
Schneider, E., 1219
Schneider, W., 1418, 1419
Schoene, M., 1099
Schoenemann, J., 1467
Schoenemann, K., 257

Schoenherr, O., 1359
Schofield, M., 303
Schofield, R.E,. 868
Schraml, C., 1047
Schraps, ---, 258
Schremmer, E., 1048
Schreyer, J., 189
Schridowitz, P., 955
Schroeder, W., 1420
Schubarth, E.L., 52
Schuett, H.W., 1140
Schuette, H.A., 820-21
Schultz, G., 259
Schulz, H., 260, 869
Schulz, R., 764
Schurer, H., 956
Schuster, C., 456, 458, 1254
Schwen, G., 1194
Schwerin von Krosigk, L., 262
Sconce, J.S., 1367
Scoville, W.G., 870-72
Sédillot, R., 1381
Seetzen, U.J., 966
Senecal, V.E., 474
Senf, F.A., 1049
Servos, John W., 379
Shanks, J., 396
Sharrer, T., 822
Shaw, P., 3, 5
Shaw, S., 688
Shée, H., 1480
Sheets, T. Jr., 1271
Shemilt, L.W., 380
Simmersbach, ---, 1050
Simpson, C.H., 1339
Singer, C., 141, 638
Sipley, L.W., 1448
Skempton, A.W., 901
Sloan, A.E., 823
Slokar, J., 204
Smith, D.C., 928
Smith, F.P., 182
Smith, J.G., 224
Smith, J.L., 66, 72
Smith, R.A., 302, 553, 571

Snow, J., 1421
Society of Apothecaries, 1422
Socquet, J.M., 1478
Sohlman, R., 440
Sonnedecker, G., 1403, 1423
Sonnemann, R., 1256
Spackman, C., 902
Spears, J.R., 667
Spence, P., 552
Spener, C.F., 984
Speter, M., 263, 608, 1141, 1156, 1286, 1368, 1449
Spiegel, L., 501
Sprengel, P.N., 17
Spronsen, J.W., 425
Sproxton, F., 1195
Squarzina, F., 325
Srbik, H.R., 1051
Stadler, G., 1036
Stander, S., 1424
Stapleton, H.E., 975
Stearns, R.P., 1112
Stein, C.M.A., 381
Steinbeck, E., 264
Steinert, O., 453
Stenger, E., 1450
Stephens, M.D., 439, 570
Stephens, T., 581-82
Stern, H.J., 957
Steuben Glass, firm, 872
Stieb, E.W., 1425
Stieda, ---, 1111
Stocking, G.W., 1468
Stoepel, K.T., 1469
Stoeri, F., 1196
Storek, J., 824
Storz, L., 1313
Stratton, H.J., 689
Strubbe, I., 142
Struve, H., 18
Stuhlmann, C.C., 183
Stutzer-Koenigsberg, ---, 1470
Stuyvenberg, J.H., 825
Suckow, G.A., 23, 237
Sudhoff, K., 1157
Suhling, L., 1382

Suler, B., 335
Svedenstjerna, E.T., 274
Swedenborg, E., 980
Swindin, N., 115, 447
Sworykin, A.A., 143
Szepczynski, S., 1314
Szokefalvi-Nagi, Z., 609, 639

Talbot, G., 613
Taube, F.W., 271
Taussig, R., 1350
Taylor, F.S., 144, 1158
Teague, W.D., 824
Te Brake, W.H., 572
Teeple, J.E., 1471
Teichowa, A., 205
Tennant, C., 1203
Tennant, E.W.D., 304, 526
Tennant, S., 1116
TerMeer, F., 495
Thackrah, C.T., 547
Thenard, L., 43
Theophilus, 698
Thépot, A., 225
Thévenot, J., 1058
Thiele, O., 1092
Thoelden, J., 976
Thomas, D.B., 1451
Thomas, J., 690-91
Thomas, P.J., 765
Thomas, R.E., 491
Thompson, D.V., 766-67
Thorpe, W.H., 874
Thurston, E.F., 305
Tillet, M., 538
Timm, B., 1159, 1340
Tooke, W., 328
Toulmin, S., 1152
Townsend, J., 338
Townson, R., 190
<u>Tradition</u>, 450
Traill, T.S., 988
Trescott, M.M., 1271
Treue, W., 145
Troitzsch, U., 265

Trommsdorf, J.B., 32
Tropke, J., 878
Trump, E.N., 1506
Tschudi, P., 527
Tulloch, T.G., 1287
Turgan, J.F., 226
Turnbull, G., 768
Turrière, E., 875
Turrill, P.L., 382
Turton, R.B., 640
Tuttle, F.J., 1528
Tuttle, W.M., Jr., 383
Tytler, J., 35

Uilkins, J.A., 313
Ulloa, D.G.J., 336
Underwood, A.J.V., 1160
United States, government, 347-48, 384-85
United States Borax & Chemical, firm, 668
Upmann, J., 1288
Urbanski, M.T., 1289
Urdang, G., 1403
Ure, A., 53
Urquhart, A.R., 1197

Van Antwerpen, F.J., 386
Van Gelder, A.P., 1290
Vauquelin, L.LW., 625-26, 718, 1481
Veatch, J.A., 652
Veneable, F.P., 1350
Ver Planck, W.E., 669
Vershofen, W., 146
Vetterli, A., 769
Viennet, O., 210
Viewig, W., 90
Vincent, F., 1052
Vivian, J.E., 404
Voeglin, W., 1257
Voelcker, H., 497
Vogel, F.M., 266
Vogel, J.H., 1341
Vogel, M., 1258
Vogel, O., 1519
Volger, W.F., 1053
Volkmann, J.J., 310

Voorn, H., 931-32
Vorce, L.O., 1272

Waeser, B., 1342, 1520-22
Wagner, J.A., 58
Wailes, R., 733
Waksman, S.A., 1426
Wall, E.J., 1452
Waller, R., 701
Wallerius, J.G., 1115
Walter, G.A., 267
Wang Ling, 184
Ward, E.R., 427, 1259, 1260, 1360
Warren, K., 306
Warren, P., 580
Warrington, C.J.S., 387-88
Wascha, O., 692
Watson, R., 22
Watts, G.B., 14
Weatherill, L., 693
Weber, G.A., 1142
Weber, H.C., 389
Weber, J.A., 645, 964, 1064
Wedge, U., 1522
Wedgwood, J., 868
Weeks, L.H., 929
Wehlte, E., 1143
Wehlte, K., 770
Weight, H., 661
Weinrich, P.H., 694
Weir, T.S., 930
Weise, J.C.G., 35
Weiss, E., 1220
Weiss, H.B., 1054
Weiss, O., 1055
Weldon, W., 1507
Welham, R.D., 1261
Wells, A.E., 1523
Welsch, F., 268
Welsh, P.C., 881
Wenland, R.T., 1369
Werner, C.J., 1056
Wescher, H., 771
Westrumb, J.F., 713, 1476
Westwater, J.W., 390

Wheatly, P., 185
White, A.H., 391
White, H.J. Jr., 1262
Whitehead, D., 471
Whittaker, C.M., 307
Whittemore, G.F., 1343
Wickersheimer, E., 573
Wiegleb, J.C., 10, 18, 21, 592, 1473
Wietschoreck, H., 1427
Wik, L., 392
Wilbert, M.I., 393
Wildenhayn ---. 590
Wiley, H.W., 1144
Wilkinson, N., 394
Willekip, J.J., 623
Williams, A.R., 1093
Williams, G.C., 404
Williams, T.I., 147, 463
Williamson, H.F., 1383
Wilson, C.H., 528
Wilson, C.M., 958
Wilson, C.W., 1315
Wilson, D., 1428
Winkelmann, H., 983
Winkler, A.F., 772
Winkler, C., 1524
Winnacker, K., 448
Witschakowski, W., 1161
Witt, O.N., 83, 148-49, 269-70
Witthoeft, H., 1057
Woat, T., 206
Wolf, G., 455
Woltereck, H., 317
Wood, R.D., 1453
Woodruff, W., 308, 959
Woodward, J. 703
Woulfe, P, 709
Wrany, A., 207
Wuest, H.M., 1429

Yamazaki, T., 186
Yang Tzu-Chiu, 187

Zart, A., 150
Zedler, J.H., 4

Zenghelis, C., 942
Ziemke, P.C., 1094
Zimmermann, P.A., 152, 153
Zincke, G.W., 6
Zorn, W., 1291
Zuman, F., 876
Zwehtkoff, P., 188

TITLE INDEX

abrasives, 1113-14
Accum, F, 405
acetylene, 1314, 1325, 1327, 1337, 1341, 1365
Achard, FK, 1105, 1110-11
Acheson, EG, 395, 406
acid, acetic, see vinegar
acid, boric, see borax
acid, hydrochloric, 1200, 1520
acid, nitric, 244, 309, 311, 316, 589,593, 1351-52, 1357, 1360
acid, pyroligneous, 905, 908
acid, sulphuric, 194, 206, 224, 236, 244, 273, 275, 298, 316, 324, 334, 385, 444-45, 558, 574-75, 588, 592, 633, 1200, 1487, 1508-24
acids, mineral (see also individual acids) 574-75, 588-89, 591-93, 604, 624
aerosols, 377
Africa (see also Egypt), 586, 1002, 1152
African Explosives & Chemical Industries, firm, 451
Agricola, G, 397, 611
agricultural chemistry (see also fertilizer), 392, 485, 882, 888, 1115-44
Air Liquid, firm, 449
al-Biruui, 682
alcohol (see also distillation, alcoholic beverages), 909, 1294
alizarin, synthetic, 1232, 1236, 1240, 1244, 1248, 1252, 1254
alkalis (see also soda, potash), 576-87, 590, 594-603, 606-09, 1198-99
alkaloids (see also individual substances), 1429
Allerly Matkel, 700
Allhusen, C, 396
al-Razi, 974
alum, 36, 190-91, 228-29, 234-35, 248, 264, 274-75, 296, 318-19, 326, 328, 337, 347, 613-22, 625-32, 634-40
American Institute of Chemical Engineers, 373, 386
American Petroleum Institute, 1378
ammonia, 120, 217, 359, 1145, 1155, 1159, 1161, 1321, 1323, 1328, 1333, 1336, 1340, 1357
Amsterdam, see Holland
anesthesia, 1421
anilin (see also dyes, synthetic), 1255, 1258

antibiotics (see also individual drugs), 1384, 1391, 1392, 1395, 1426
apparatus, 349, 1145-61
aqua Fortis see nitric acid
Asia, central, 158, 179-80, 719, 763, 974
Asia, southeast, 181
Aspedin, J, 890
asphalt, 352
aspirin, 1394
Atlas Cement, firm, 452
Auersperg, firm, 206
Austria, 191, 193-94, 196-202, 204, 1024, 1100-01, 1220, 1517

Babylonia, see Mesopotamia
Badische Anilin and Soda-Fabrik (BASF), firm see BASF
Baekeland, LH, 407-08
Baeyer, A, 398
Baist, L, 409
Banting, F, 1405
barilla, see alkali
barium compounds, 45
BASF, firm, 93, 132, 422, 453-59, 1155, 1184, 1193, 1196, 1244, 1296, 1335, 1340, 1359
Baumé, A, 219
Bavaria, 833, 1020
Bayer, firm, 418, 461, 1364, 1394
Becher, JJ, 932, 1216
Beckmann, J, 983
Behring, E, 1398
Belgium, 316, 746, 831, 863, 1002
Bergius, F, 410
Berthollet, CL, 1204, 1316
Berzelius, J, 1368
beverages, see food and drink, brewing, wine
beverages, alcoholic, 780, 785, 790, 812
beverages, nonalcoholic, 809, 817
Bickford, W, 1278
Bingham, R, 1444
Birkland-Eyde process, 1358
Black, J, 1472
bleaching, 1-153 passim, 224, 275, 294, 1198-1210, 1526
bleaching powder, 1207-08
Böttger, JF, 677-78, 681, 685
Bohemia, see Czechoslovakia

borax, 1-153 passim, 185, 309, 641-69
Bosch, C, 403, 410-11, 1329
Bradley, CS, 395
Brahe, Tycho, 923
Brand, H, 940
Brand, WT, 1422
bread, 801, 807, 824
brewing, 11, 36, 301, 779, 781, 784, 787, 790, 793, 798, 803-4, 813-14, 819
brickmaking, 671, 675, 687, 887
Brockedon, W, 957
bromine, 468
Brüning, A, 400
Bruges, PT de, 1472
Brunck, H, 397, 403, 413
Bruner, Mond & Co., 305, 462
Bruton, W, 1303

calcium carbide, 1331, 1341, 1345-47, 1350
calcium nitrate, 1355
California, 652, 654-56, 659-62, 665, 667-69
Calvert, FG, 1259
Cameralists, works by, 6, 19, 20, 265
Canada, 377, 380, 387-88
canning, see food preservation
"carborundum" (silicon carbide), 406
Caro, H, 397, 403, 414
cartels, 205, 260, 1332, 1463, 1467-69
Castner, HY, 395, 415
Castner Kellner, firm, 463
catalysis (see also, synthesis, high temperature and pressure), 1368
Cavour, C, 323, 1132
celluloid, 1177-78, 1188-89, 1195
Celluloid Co., firm, 1189
cellulose nitrate, 56, 1279-80, 1293, 1289, 1528
cement, 452, 523, 887-88, 890-91, 893-94, 896-98, 900-02
ceramics (see also porcelin), 45, 166, 176, 187, 275, 277, 301, 670-94
chagrin, 44
Chance Bros., firm, 464-65
charcoal, 14
"chemical engineering," 92, 100, 116, 156, 168, 195, 257, 320, 362, 368, 372, 379-80, 389-91
chemical hazards, 534-73

chemical industries, location of, 103, 132, 243, 256
chemical technologies, classification of, 6, 21, 25, 30, 32, 41-2, 46, 49
chemical warfare, 120, 353, 371, 377, 937, 1273, 1326, 1343
Cheshire, 977, 999
Cheremond, 965
Chiddingford, Surrey, 844
China, 154, 159, 164-67, 169-76, 180-85, 187-8, 647, 679, 682, 920, 939
chlorine, 224, 275, 294, 1201-05, 1269, 1271-72, 1316, 1319-20, 1326, 1330, 1333, 1335, 1338, 1361
chloroform, 1421
chromium, compounds, 718
CIBA, firm, 466
Clegg, S, 1311
coal, 14, 22, 49, 56, 134, 169, 271, 273-74, 295-96, 301, 1211-20, 1292, 1298, 1303, 1305, 1310
coal, fluidization, 1218
coal oil, 1216
coal tar (see also dyes, synthetic), 1213-15, 1259
cocaine, 1404
Cochran, Archibald, see Dundonald
Collison, J, 1472
Commercial Solvents, firm, 467
company histories, 88, 251, 287, 292, 366, 375, 449-533, 1114, 1290
Compositiones ed tinguenda, 747
Consolidirte Alkaliwerk Westeregeln, firm, 468
Cookson, I, 1472
copperas, see vitriol
cortisone, 1402
Cottrell, FG, 416
Cowles, AH, 395
Crossfield, firm, 469
Curadau, FA, 628
cyanogen, 1322
Czechoslovakia/Bohemia, 189, 203, 205-7, 867, 876

dacron, 1162
Daguerre, LJM, 1439, 1445, 1449, 1453
Davis, GE, 115, 168
Deacon, H, 396
DeBrisay, KW, 1002
Delius, F, 1147
Demachy, JF, 1156

Denmark, 923
Descroizilles, FAH, 1146, 1316
detergents, 1099
Deutschen Solvay-Werke, firm, 470
Dickson, JT, 1162
distillation, 1-153 passim, 1148-49, 1151-53, 1156-58, 1160
Dixon, G, 1310
Dennis, M, 401
Domagk, G, 1391, 1415
Doubleday, T, 277
Dow, HH, 401
Dow, firm, 471
dry cleaning, 46, 700
Duisberg, C, 420, 1388
Dumas, JB, 1418
Dundonald, 9th Earl of (Archibald Cochrane), 401, 421, 1214-15
DuPont, firm, 449, 472-74, 1168, 1173, 1186
dyes and pigments, traditional (see also lead and mercury compounds of, verdegris), 1-153 passim, 169, 267, 275, 294, 527, 695-772, 1526-27
dyes, synthetic, 226, 299, 307, 346, 351, 369, 427, 430, 443, 1221-62

Egypt, 594, 724, 755-56, 804, 860, 898, 902, 961-62, 970, 1060, 1066
Ehrlich, P, 1397
Electrochemisk a/g, firm, 475
electrochemistry, 86, 145, 395, 476, 491, 511, 1263-72
enameling, 672
encyclopedias, 8, 12, 13, 16, 17, 33, 35, 47
England (see also Great Britain), 614-15, 636, 640, 693, 733, 754, 765, 813, 839, 843, 849, 853, 863, 865, 874, 985, 998, 1017-18, 1307, 1510
explosives (see also pyrotechnics), 353, 377, 440-42, 451, 510, 1273-91
Exxon, firm, 476

Fahlberg, C, 1401
Farm Chemurgic Council (US), 392, 1137
Feldman, WH, 1386
fertilizer (see also agricultural chemistry, phosphorus) 882, 888, 1116, 1118, 1121, 1127, 1129-30, 1141, 1143, 1355
fibres, artificial, 1166-68, 1171, 1173-74, 1181, 1184-86, 1191, 1197

Firestone, firm, 477
fireworks, see pyrotechnics
Florida, 1120
flour, see bread
food & drink, 226, 773-825
food preservation, 773-75, 777, 782-83, 788-89, 796, 800, 802, 808, 811, 1154
Ford, H, 392
Ford, JB, 827
Fourcroy, AF, 222, 1286, 1481
Fox Talbot, WH, 1439, 1442-43, 1445, 1447
France, 208-26, 545, 625, 628, 630, 674, 727, 771, 793, 828, 841, 865, 872, 885, 971, 1002, 1028, 1043, 1069, 1073, 1082-84, 1086-87, 1089, 1091, 1096, 1107, 1307, 1484
Frank, A, 397
Frasch, H, 1377
Frauenhofer, J, 832
Fresnel, L, 1498
Fry, J, 1472
Fuch, 1344
fuels, (see also coal, gas), 378
fuels, rocket, 1293
fungicides, see insecticides

galilith, 1166
galvanism, 33
galvanoplastics, 1270
gas, heating and illuminating, 49, 54-5, 134, 226, 275, 421, 434, 1298-1315, 1370
Gas Light & Coke Co., firm, 478
gases, industrial, 1316-43
gasometer, 1150
Geigy, firm, 479
General Chemical, firm, 480
Genoa, Italy, 635
Germany, 82, 84, 88, 227-70, 374, 684, 750, 857, 866, 964, 1024, 1039, 1050, 1078, 1222-23, 1237-38, 1256, 1278, 1291-92, 1309, 1338, 1364, 1390, 1417, 1420, 1427, 1456, 1462, 1467-69, 1494, 1499
Gilbert, JH 1127
Giulini, G, 403
Glasgow, 1500
glass, 1, 14, 45, 117, 157, 176, 189, 201, 226, 301, 464-65, 499, 507, 514, 826-76
Glauber, JR, 138, 397, 424-5, 937

Glover, J, 538
glue, 14, 44
Goldschmidt, firm, 481
Goodyear, C, 946-47, 959
Goodyear, firm, 482-83
Gossage, W, 396
Graebe, C, 1236
graphite, 406, 1263
Graselli, ER, 401
Gravenhorst, firm, 229, 621, 708
Great Britain (see also England, Scotland), 276-308, 554-55, 596, 599, 673, 675, 690-91, 768, 786, 800, 891, 904-5, 959, 1139, 1171, 1225, 1229, 1234, 1260-61, 1273, 1282, 1287, 1385, 1406, 1425, 1482 1493, 1501, 1516
Greek fire, 936, 939, 941-42
Griesheim, firm, 409, 484-486
Griess, P, 397, 426-28
guano, 336
Guimet, JB, 719, 763
gun cotton (cellulose nitrate), 56
gun powder, 1-153 passim, 175-76, 184, 245, 296, 347, 394, 935, 939, 1065-66, 1069
gutta percha, 945-46
gypsum, 1-153 passim, 338, 895, 898

Haber, F, 398, 1329
Hales, S, 1305
Hall, CM, 395
Hall, Tirol, 970, 1013, 1037, 1045
Halle, Prussia, 1006-07
Hallein, Austria, 970, 983, 1021, 1036
Hannover, 1457
Harbison-Walker, firm, 487
Hare, R, 1346
Harmant, PG, 1430
Harz region, Germany, 1038
Hasslacher, J, 401
Hazard, R, 429
helium, 1334
Henkel, firm, 488
Hermann & Sohn, firm, 1494
Herschel, J, 1445, 1447
Hesse, 623, 1026, 1055
Hinsberg, O, 1388
Hinshaw, HC, 1386

Hoechst, firm, 125, 448, 489-90, 1182-83, 1396-98, 1438, 1513
Hofmann, AW, 397-99, 430, 1249
Hofmann, F, 1364
Holker, J, 1509
Holland, 2, 27, 309-15, 317, 931-32, 966, 1210
Hooker, firm, 491
hormones, synthetic, 1396
Huls, firm, 492
Hungary, 190, 192, 585, 609, 639
Huygens, C, 1150
Hyatt, JW, 1177-78, 1188-89
hydrochloric acid, see acid
hydrogen, 1318
hydrogenation, 1292, 1296

ice cream, 806, 821
IG Farbenindustrie, firm, 449, 494-96, 1362
Illinois, 1033
illumination, see gas, heating and illuminating
Imperial Chemical Industries, firm, 449, 493, 1162
India, 156, 160-61, 177-78, 646, 721-22, 740, 812, 920, 992, 1058, 1081, 1090
Indians, American, 98, 927
indigo (dye), 94, 769, 1221, 1225-26, 1235, 1244, 1250
industrial chemistry, exhibitions of, 59-91, 212
industrial chemistry, general history, 92-153
industrial chemistry, prehistoric, 98, 135
industrial research, 1222
insecticides and and fungicides, 353, 1124, 1133-34, 1138
insulin, 1405
iodine, 601, 1348-49
Ipatieff, VN, 431
Ireland, 874
Iriny, 1291
Italy, 92, 320-25, 615, 618, 634, 644, 650, 657-58, 664, 666, 829, 839, 1132

Jamestown, Virginia, 334
Japan, 155, 157, 162, 163, 168, 186, 920, 1002
Jefferson, T, 350
Johnson, IC, 900

Kalle, firm, 497
Kammerer, JF, 1291
Kanawha Valley, W. Virginia, 357, 360, 987, 991, 1014

Kansas, 1052
Keir, J, 303, 432, 1472, 1502
kelp, see alkali, potash
Kentucky, 1068, 1076, 1094
Klipstein, A, 401
Klipstein, EC, 401
Klipstein, KH, 1229
Knietsch, R, 403
Knoll, A, 403
Kolbe, H, 1413
Koller, C, 1404
Kosta, firm, 499
Kränzlein, 1182
Kuhlmann, firm, 449, 500
Kunheim, firm, 238, 501
Kurtz, A, 396

lacquer, 741, 746
Lampadivs, WA, 1141, 1524
Lancashire, 302
lapis lazuli, see ultramarine
Latin America, 336, 342, 648, 653, 927, 943, 1002, 1067, 1070-71, 1074-75, 1077, 1079-80, 1085, 1102, 1109, 1112
Lautemann, E, 1413
Lavoisier, A, 1091
Lawes, JB, 1127
lead, compounds of, 1-153 passim, 155, 187, 191, 199, 245, 248, 311, 327, 533, 739
leather, 9, 25, 54, 277, 301, 394, 738, 757, 877-81
Leblanc, N, 218, 397, 433, 1483-84, 1489-92, 1505
Lebon, P, 434, 1303, 1305
Leonardo da Vinci, 1157
Lewis, GT, 401
Lewis, WK, 435
Liebermann, CT, 1236
Liebig, J, 250, 418, 1125, 1135, 1138
Liege, 612, 619, 965
Lille, 604, 1515
lime, 1-153 passim, 169, 236, 311, 347-48, 882-89, 892, 896, 899, 902, 1116, 1118
linoleum, 1175
Lorraine, 839
Losh, W, 277
Lucius, E, 400
Lüneburg, 1025, 1049, 1053, 1057

Lunge, G, 397, 436

Mackintosh, C, 274, 956-57, 1528
Makkus, W, 1355
Mallinckrodt, E, 401
Mansfield, CB, 1259, 1360
Mapes, JJ, 401
Marseilles, 1096
Massachusetts, 583-84
Massachusetts Institute of Technology, 368, 379, 389, 404, 435
matches, 1274-76, 1278, 1281, 1291
Mathieson, firm, 502
Matthes & Weber, firm, 503
mauve (dye), 1227
McBee, ET, 1361
Meissen, 692
Meister, W, 400
Meister, Lucius & Bruening, firm, 505
Mercer, J, 1525
mercerizing, 1525
Merck, E, 146
Merck, firm, 506
Mercury, compounds of, 1-153 passim, 162, 187, 230, 309, 311, 766
Merrill, J, 1374
Mesopotamia, ancient, 637, 724, 803, 810, 819, 860-61, 895, 1098, 1153
Michaels, F, 1454
Michigan Alkali, firm, 507
Missouri, 687
Mittasch, A, 403
Moissan, H, 398, 1350
Mojave desert, California, 382
molasses, 816
Mond, A, 437
Mond, L, 290, 437-38, 1332
Monsanto, firm, 508
Montecatini, firm, 449, 509
morphine, 1418
mosaic gold (aurum mosaicum) 709, 758
Moulton, S, 959
Murano (Venetia), 829, 842, 863
Murano glass, 226
Murdoch, W, 1303, 1305-6, 1308, 1310-11

Murray, Sir J, 1123
Muspratt, JS, 295, 396, 439, 536, 570

natron, 594, 756
Near East, 839, 917, 930, 939, 973-75, 1098
Nef, John V. 843
New Jersey, 1054
New Mexico, 1464
New York, state and city, 994, 1056, 1270-71, 1300
Newcastle, England, 277, 279, 1487
niello, 743
Niepce, N, 1437, 1443
nitrates, synthetic (see also ammonia), 340, 1121, 1351, 1353, 1356-59
nitric acid, see acid
nitrogen, fixation of, 1321-23, 1327, 1332, 1335, 1340, 1342, 1351, 1353, 1356, 1358-59
Nobel, A, 440-42
Nobel, firm, 510
Norsk Hydro, firm, 340, 511
North America, 579, 582-84, 598, 600
North Carolina, 1340
Norway, 1351, 1359
nylon, 1168, 1173

oil cloth, 1528
oils, vegetable, 352, 815, 904, 906-7, 911-12
oleomargarine, 820, 825
Ostwald, W, 1356
oxygen, 1317, 1329

Pacific Coast Borax, firm, 512
paper, 1-153 passim, 159, 165, 169, 277, 296, 394, 913-32
parchment, 14, 44
Paris, 889
Parkes, A, 1177, 1188-89
Pastor, PH, 549
patents, 139, 151, 362
Patison, J, 1217
Payen, A, 399
Pelletier, PJ, 1418
Pelouze, TJ, 1279
penicillan, 353, 1389, 1391, 1395, 1412, 1428
Pennsylvania Salt, firm, 513
Perkin, WH, 399, 443, 1225, 1228, 1242, 1251, 1257

Perret, M, 1508
petrochemicals, 521, 1361, 1369, 1371, 1378, 1383
petroleum, 76, 191, 374, 378, 475, 498, 521-22, 936, 1292, 1295, 1370-83
pharmaceuticals, 1-91 passim, 167, 177, 219, 251, 258, 309, 778, 1384-1429
phenolformaldehyde plastics, 1170
Philadelphia, Pennsylvania, 378, 393, 1404
phosgene, 1324
phosphores & phosphates, 323, 940, 1099, 1120, 1123, 1129, 1132, 1136, 1141
photography, 134, 1430-53
pigments, see dyes and pigments, lead, mercury, verdigris
Pilkington, firm, 514
Pittsburgh Plate Glass, firm, 827
plastics, artificial, 918, 952-53, 1162-65, 1169-70, 1172, 1175, 1177-78, 1180, 1182-84, 1187-90, 1192-96
plastics, natural, 944, 946, 1176, 1179
Poland, 334
polyethylene, 1162, 1180
Pompei, 732
porcelain, 14, 170, 172, 187, 275, 677, 683-85, 692, 856
potash (potassium salts), 36, 39, 45, 189, 191, 234, 248, 271, 275, 326, 328, 347, 498, 531, 577, 579-84, 587, 590, 595, 597-98, 600, 603, 606-9, 1010, 1125, 1130-31, 1434-71
Pott, JH, 608
pottery, see ceramics
Prestonpans, Scotland, 273
Prussian blue (Berlin blue), 1-153 passim, 154, 234-35, 244, 248, 355, 702-4, 712-13, 716, 726, 734, 1322
pyrotechnics (see also explosives), 1, 3, 175, 184, 353, 933-42, 1067

Queeny, JF, 401

Raschig, F, 403
rayon, 1167, 1174, 1181, 1186, 1197
Reichenhall, Bavaria, 984, 1005, 1021, 1027, 1048
Reiman, L, 403
refractories, 676, 680
Remsen, I, 1401
Rhineland, 883
Riedel, firm, 238
Rockefeller, JP, 1377
Roebuck, J, 298, 303, 445

Roger of Helmarshausen, 698
Rohm & Haas, firm, 516
Rome, ancient, 735, 738, 740, 882, 892
Rosengarten, DG, 401
Rothamsted Agricultural Experiment Station, 1127
Rotheim, E, 1406
rubber, 226, 308, 383, 477, 483, 492, 530, 943-46, 948-59, 1362, 1364, 1366, 1528
rubber, synthetic, 953-54, 958, 1184
ruby glass, 855
Russia, 326-33, 335, 580, 595, 1378

Saar, 627
saccharine, 1401
Saint-Gobain, firm, 226, 517-19, 834, 864
St. Helens, England, 279
St. Rollox, Scotland, 289, 304, 525-26
sal ammoniac (ammonium chloride), 39, 158, 167, 179, 219, 230, 235, 316, 318, 960-75, 1481
Salins, France, 1040
salt (sodium chloride), 36, 163-64, 169, 171, 174, 176, 188, 190, 192, 228, 272, 275, 311, 326, 328, 337-38, 347-48, 355, 385, 976-1057
saltpeter (potassium nitrate), 36, 39, 176, 191-92, 202, 230, 233, 244, 248, 264, 296, 326, 328, 337-38, 347, 355, 935, 1058-66, 1068-69, 1072-73, 1076, 1081-84, 1086-87, 1089-94
saltpeter, Chili (sodium nitrate), 1067, 1070-71, 1074-75, 1077-80, 1085, 1088, 1092, 1354
salvarsan, 1397
Salzkammergut, Austria, 989, 1047, 1051
Saxony, 1009
Schauberger, O, 982
Schönebeck, Prussia, 1001
Schonbein, CF, 250, 1189, 1279
Schott, firm, 520
Schulze, JH, 1450
Scotland, 596, 601, 631, 1500
Searles Lake (California), 656, 1471
Seille Valley, France, 990, 1042
Sertürner, FW, 1418
Shannon, A, 1472
Shaw, Peter, 282
Sheffield, WE, 1512
Shell (Netherlands), firm, 521

Shell (United States), firm, 522
shellac, 1179
Shirley, T, 1311
Silesia, 253, 624, 1519
Skanska Cement, firm, 523
Skey, S, 298
smalt (cobalt pigment), 1-91 passim, 94, 189, 245, 248, 311, 706-7, 711, 742, 772
Smith, RA, 532-33, 535
soap, 1-153 passim, 94, 187, 226, 275, 469, 1095-98
Sobrero, A, 1279
soda, 45, 183, 190, 218, 224-25, 229, 268, 275-80, 286, 290-91, 294, 297, 306, 338, 385, 429, 432-33, 462-64, 470, 524, 529, 578, 585-86, 594, 596, 599, 602, 1206, 1472-1507
soda, caustic (sodium hydroxide), 1268-69, 1271-72, 1499
Solfaterra, Italy, 319
Solvay, E, 399, 1485, 1498, 1506
Solvay, firm, 524
South America, see Latin America
Spain, 337-41, 988, 1106
Spence, P, 396
Spill, D, 1189
Staffordshire, 688, 691, 693
starch, 14, 36, 275
Stassfurt, 995, 1010, 1454, 1458-9, 1463
Staudinger, H, 1182
Stephens, T, 607
Stolz, F, 1396
streptomycin, 1386
sugar, 9, 35, 169, 226, 296, 1100-12
sulfonamid, 1391, 1415
sulphur, 169, 189, 192, 194, 275, 309, 318, 325, 336, 347, 423, 612-13, 624, 1133, 1508
sulphuric acid, see acid
Sussex, 905
Sweden, 597, 613, 620, 883, 1072
Swindin, N, 447
Switzerland, 343-46, 892
synthesis, high temperature and pressure, 431, 1145, 1147, 1155, 1159, 1161, 1292, 1295-97
Szechuan, province, China, 171, 174

tanning, see leather
Tennant, C, 526

Tennant, firm, 525-26, 1207
textiles (see also dyes), 9, 92, 96, 176, 348, 481, 751, 761, 765, 768, 1525-28
Theophilis, 855
Tibet, 643, 647, 651
Tolfa, Italy, 319, 618, 628, 634
transport phenomena, 92, 97
Transylvania, 1004
Trommsdorff, JB, 146, 1408
Tschirnhaus, EW, 667, 685
Tschudi, firm, 527
Tuscany, 644, 649-50, 657-58, 664, 666

Ulstad, P, 1157
ultramarine, 719, 763
Unilever, firm, 528
unit operations, 115, 118, 145, 349, 363, 368
United Alkali, firm, 276, 529
United Kingdom, see Great Britain
United States, 76, 91, 347-79, 381-87, 390-94, 603, 606-7, 632, 675, 686, 689, 725, 744, 777-78, 781, 790, 815, 818, 822, 835, 858, 870-71, 875, 880-81, 894, 899, 906, 928-29, 952, 958, 1035, 1041, 1061, 1121, 1138, 1230, 1267, 1271, 1290, 1292, 1294-95 1299, 1312-13, 1315, 1343, 1353, 1370-72,, 1376, 1383, 1399, 1412, 1423, 1440, 1464-65, 1470-71, 1485, 1504, 1506, 1522-23
United States, Dept. of Agriculture, 361, 1128, 1142
United States Rubber, firm, 530
uranium, 498
urine, 903

Vauquelin, LW, 222, 1286
Venice, 863, 875, 1015, 1030
verdegris, 94, 244-45, 717
Vershofen, W, 1417
Vienna, 967, 1518
vinegar, 605, 905, 910
Virginia, 845
vitamins, 817
vitriol (iron and copper sulphates), 36, 189, 191-92, 194, 230, 233, 248, 264, 274, 296, 318, 326-28, 336, 347-48, 610-13, 616, 618-19, 623-24, 633
Vorster & Grueneberg, firm, 531

Waksman, LS, 1386

Walker, J, 1281, 1291
Walton, F, 1176
Warner, LC, 401
WASAG, firm, 532
Washburn, FS, 401
water glass (sodium silicate), 1344
Watt, J, 1472
wax, 14
Weald, The (England), 852
Weber, CO, 951
Wedgewood, J, 868
Wenod, J, 1157
Westphelia, 1012, 1029
Wetherill, firm, 533
Whinfield, JR, 1162
Widnes, England, 284
Wieliezka, Poland, 1003, 1016, 1019
Wiley, HW, 778
Willson, TL, 1340, 1350
Wine, 3, 776, 792, 805
Winkler, C, 397, 1524
Winkler, F, 403, 1218
Winsor, A, 1307
Winthrop, J, 401
Wisconsin, Agricultural Experiment Station, 808
Wisconsin, University of, 97, 368
woad (dye), 37, 733, 745
Wöhler, F, 250
World War I, 108, 120, 288, 875, 1121, 1327, 1343
World War II, 353, 377, 383, 490, 952, 958
Württemberg, 1000, 1011

yeast, 797
Young, W, 680

Zollverein, 239